THE COMPLETE GUIDE TO

TO

MEMORY

THE COMPLETE GUIDE TO MEMORY

THE SCIENCE OF STRENGTHENING YOUR MIND

RICHARD RESTAK, MD

Skyhorse Publishing

Skyhorse Publishing books may be purchased in bulk at special discounts for sales promotion, corporate gifts, fund-raising, or educational purposes. Special editions can also be created to specifications. For details, contact the Special Sales Department, Skyhorse Publishing, 307 West 36th Street, 11th Floor, New York, NY 10018 or info@skyhorsepublishing.com.

Skyhorse® and Skyhorse Publishing® are registered trademarks of Skyhorse Publishing, Inc.®, a Delaware corporation.

Visit our website at www.skyhorsepublishing.com

10 9 8 7 6 5 4 3 2

Library of Congress Cataloging-in-Publication Data is available on file.

Hardcover ISBN: 978-1-5107-7027-0
Ebook ISBN: 978-1-5107-7028-7

Cover design by Mona Lin and David Ter-Avanesyan
Cover illustration by Getty Images

Printed in the United States of America

To the beloved memory of my sister, Louise Restak Ollstein

Contents

CHAPTER I

WHY SHOULD I CARE ABOUT MY MEMORY?

HOW COMMON ARE MEMORY WORRIES?

There are many reasons to care about your memory. Consider these: the development of a superpower memory enhances attention, focus, abstraction, naming, spatial visualization, verbal facility, language, and word acquisition. In a phrase, memory is the key to brain enhancement.

In America today, anyone over fifty lives in dread of the Big A—Alzheimer's disease. Small social gatherings (dinner, cocktail parties, etc.) take on the atmosphere of a segment from NPR's weekly quiz show "Wait Wait . . . Don't Tell Me." That's the one where guests vie with each other in intense competitions to be the first to come up with the names of such things as the actor playing a role in the latest mini-series everybody is binging on. Almost inevitably, someone will pull out a cellphone to check the accuracy of the person who responded first. Quick, quicker, quickest lest others suspect you of coming down with the initial symptoms of the Big A.

Although Alzheimer's disease is not nearly as common as many people fear, nevertheless worries about perceived memory lapses are increasingly expressed to friends. They are also the most common complaint that persons over fifty-five years of age bring to their doctors. Such memory concerns are often unjustified and arouse needless anxiety. This widespread anxiety has helped create a national pre-occupation with memory and signs of memory failure. One of the reasons for this panic is the confusion in many people's minds about how we form memories.

Try to remember something that happened to you earlier today. It doesn't have to be anything special—any ordinary event will do just fine. Now consider how that memory came about.

At my request, you recovered a memory for something that you probably would have not thought about, if I hadn't prompted you to recall it, and you hadn't made the effort to retrieve it.

Reduced to its essentials, memory involves re-experiencing something from the past in the form of a recollection. Operationally, memories are the end products of our efforts in the present to recover information that is stored in our brain.

Memories—like dreams and acts of the imagination—vary from one person to another. My memories are distinctly different from your memories based on our personal life experiences.

Memory also differs from pictures or videos of events from the past. While these technologically based versions of the past can serve as memory stimulators, they are not themselves memories.

IS MY MEMORY FUNCTIONING NORMALLY?

In order to answer that question, let's first address the idea of normal memory function. Below are several questions, which I invite you to check as involving either (a) an example of a perfectly acceptable memory or (b) perhaps the beginning of a potentially serious memory problem.

Question A. "After scanning the newspapers and selecting several bargains I drove to the mall, parked my car and went into the store selling these bargains. When I came out I couldn't remember where I had parked, and I had difficulty finding my car. It took me several minutes to locate it in the crowded parking lot." a or b?

Question B. "After exiting a store where I had gone to purchase several items that were on sale, I came out and couldn't remember whether I drove to the shopping center or someone dropped me off there." a or b?

Question C. "I can't remember the names of my grandchildren." a or b?

Question D. "When I get together with my sister, she and I have very different memories of events we both participated in as children." a or b?

Question E. "I used to play a pretty good game of bridge, but now I'm messing up. I can't remember what cards have already been played. Nobody now wants to partner with me." a or b?

Question F. "While driving home from the office, I took the wrong exit. I have never done that before." a or b?

Question G. "I have increasing difficulty remembering names." a or b?

Question H. "Before I go into the food market, I have to write everything down or I will forget one or two items. Yet in my work as a professional actor I have no problem, even in my eighties, studying a script over a weekend and on Monday engaging in an early rehearsal without referencing the prompt script. How can that be?" a or b?

Question I. "I packed my car ready for a trip, but when I got in the car, I had no idea where I was planning to go." a or b?

Answer A. The answer to this is *(a)* perfectly normal forgetting. At the time you drove to the mall, you most likely were mentally preoccupied with the bargains that you were intending to purchase. Your concentration was far removed from such quotidian concerns as where you will park your car. You simply slipped into the first available parking place closest to the store and rushed inside and started shopping. Since you were not at all concentrating on the location of the parking place you selected, you couldn't remember its location when you came out of the store. As Samuel Johnson stated over two hundred years ago, "The art of memory is the art of attention." The attention that Johnson referred to is an *internal* attention: riveting your mental powers on a single external object. Lacking attention to the parking place, you could only form an imperfect memory, or in this specific instance, no memory at all. We will have a lot to say about attention later.

Answer B. The correct answer here is (*b*): Perhaps a potentially early and serious impairment in normal memory functioning. In this example, instead of having difficulty remembering one item out of many (where you parked) you lost the ability to recall how you even arrived at the mall in the first place along with all the interactions that would invariably take place along the way if you were driven there (talking with the driver, listening to the radio, etc.) or on public transportation (choosing a seat, looking out the window, etc.).

Answer C. (a) Remembering the names of one's grandchildren is closely aligned with the interest you have in your grandchildren, and children in general. *Interest* is the underpinning for the attention that Samuel Johnson referred to. We rarely pay attention to things that do not interest us. Although not being interested in one's grandchildren, coupled with failure to remember their names, perhaps carries implications for potential intrafamilial friction, it is not necessarily a sign of an impaired memory.

Answer D. Siblings, even identical twins, do not share the exact same experiences. Differences in age, gender, interaction with adults, especially relatives—all of these can lead to distinctly different experiences and therefore different memories. The answer is (a).

Answer E. (b) Failure to perform secondary to forgetting overlearned procedures can be worrisome. The longer we have been doing something, the less likely we will forget it. So how much of a change does this represent? If you were only a middling player, when at your best, your current decreased performance probably doesn't qualify as a major memory concern. Perhaps you have simply become bored with the game or the players? The key question is: How much of a change does this represent? If you were a highly proficient player and now nobody wants to play with you, your memory needs further investigation.

Answer F. (a) Driving involves what neuroscientists refer to as *procedural memory*. After a task is carried out a sufficient number of times, it is no longer necessary to pay conscious attention when doing it. Driving from home to work and back again established over time a procedural memory, which includes when to exit. This is automated within the brain in a specific network. This procedural memory can be overwritten by daydreaming, lack of focus, or poor attention. This also explains *over practice effects*. Athletes have learned from experience that "trying harder" can actually interfere with their performance. In this instance the learned procedural memory, which had been established on the basis of hundreds or even thousands of hours of practice, is overwritten by a conscious motor program which had been previously automated. This is also the basis for "choking": An athlete overrides his own procedural memory program with anxiety about his performance. As can be seen from this example of missing the exit, different aspects of memory can come into conflict with each other.

Answer G. (a) This is probably the most common complaint that people of all ages express about their memory. Although they fear that they may be coming down with dementia, it is actually a quite common phenomena that fits well within the category of normal functioning. If you think about it for a moment, one name can easily be substituted for another: I'm Richard Restak, but I could just as easily have been named David Restak or Justin Restak, or even something fancy such as Sebastian Restak. The point is that there is no necessary connection between a name and a face. That's why names are so hard to remember. As we will see later, there are methods for linking faces with names. With a bit of practice, you can learn to remember the names of dozens of people and rarely experience a "senior moment": drawing a blank when trying to come up with one person's name in order to introduce them to someone else.

The second principle of memory operation after *attention* involves *creating meaning*. A name is not meaningful unless you come up with a link between that name and a vivid picture or auditory association. This is the science of mnemonics: The use of a pattern of numbers, letters, images, or associations to assist in remembering something. Mnemonics can be traced back to the Middle Ages and even earlier. We'll cover some of this early history in Chapter II.

Answer H. That actor's questions foreshadow many of the themes we will be covering in this book: Concentration, motivation, deliberation, vivid images, and more. The first part of his question about the loss of memory for grocery items is (a): Within the expected memory performance of an eighty-year-old. It becomes increasingly difficult as we age to compose shopping lists and later remember them without writing them down ahead of time. There are methods that can be used to combat this, as we will discuss in later chapters.

The second question posed by the actor is intriguing. Reading a new script and memorizing its contents has become second nature to the aging actor. It is now an autonomous process. So when he sits down to read his script, a different part of his brain is involved in memorizing the character's dialogue as compared to how his brain operates when he sits down to compose a grocery list. In the latter situation he is subject to all of the vicissitudes of memory characteristic of an eighty-year-old man. This dichotomy between his theater experience and his daily life was not consciously chosen, but resulted from his script-learning over many years in the theater. This is not inevitable among aging actors, incidentally. In later life many of them give up the theater in favor of appearing in movies because the memory burden is less challenging. But the stage doesn't necessarily have to be abandoned. One internationally acclaimed film star in his eighties wears an earbud through which prompts are conveyed when he performs on stage. Presumably, he requires this because he can no longer depend on a once powerful memory enabling him to rapidly learn a script—a process in the acting profession known as a "quick study."

When I asked a professional actor in her forties regarding the dilemma of the eighty-year-old who could remember scripts, but couldn't remember grocery lists, she told me: "My best guess is that when he attempts to learn lines he is not simply memorizing words. While he is reading the script of the character that he will be playing, he anticipates taking certain actions and identifies the motivation of the character who says specific words. But the older actor is embodying not just words but thoughts, so that the words of the script are processed as expressions of the character's personality. There is an element of 'muscle memory' involved too in learning a script. It is utterly different from remembering a shopping list. When we speak words on stage we use our whole body and mind to connect with the motivation and intention of the character we play. The process of learning a script is a purposely intense one, very concentrated and deliberate. The beat and rhythm of the words connect with brain areas well beyond those where language lives."

Answer I. As unbelievable as this experience may seem, the answer in this *individual instance* is (a). This Clark Griswold-like experience was described to me by an overworked, stressed patient, who had reluctantly agreed to use up some of his long accumulated vacation time. Although he had made most all of the vacation arrangements himself (Ellen did the rest), his contribution occurred during anxiety and mild panic episodes. A full medical work-up including full memory investigation revealed normal function. So you can put this rather bizarre-sounding true example in the list of normal functioning but only as an illustration that every memory complaint must be put in the context of the individual voicing the complaint.

CAN I TRUST MY MEMORY?

Whether you can trust your memory depends quite a bit on how well you train your memory.

Lesser forms of amnesia (forgetfulness) are a normal part of life for the vast majority of people. Every time you try, but fail to remember something you once knew, you are experiencing a mild form of amnesia. Thus, amnesia isn't a rare memory disorder. Do you consider yourself an exception? Let's test that hypothesis.

What did you have for lunch today? No problem remembering that, right? How about your lunch one month ago from today? If you can remember that, I suspect something special must have occurred that day. Let's got back a month further: How about your memory for your lunch two months ago from today? (Some mnemonists can remember that). Once again, unless the lunch two months ago corresponded to some special event or special happening, it's most unusual that you would be able to remember it.

The mini-experiment you just participated in establishes an important point.

When challenged, most of us will remember far less than we think we can. But our memory will greatly improve if something eventful or emotionally arousing is associated with what we are trying to remember. Memory is pegged to emotional significance.

Three processes underlie the formation of a memory: *encoding, storage,* and *retrieval.* An example of encoding error: Suppose you were preoccupied or daydreaming when I introduced you to someone new. As a result, you are unlikely to remember the name. Because of your lack of attention the memory was not established or encoded, and therefore you couldn't come up with it later. Literally, there was no memory.

A retrieval error involves successfully encoding a piece of information, storing it, and yet being unable to retrieve it. In such a case, the information may have been encoded, but you are not able to access it.

One distinction between an encoding error and a retrieval error involves your ability to come up with the information under different circumstances. Since the retrieval errors are quite common, most tests are designed for multiple-choice. The test is easier to administer and score and you'll do better in multiple-choice because you can recognize an answer correctly that you may not have been able to come up with in response to a direct question.

Of the three stages of memory, the first—encoding—is the critical one. Many of the suggestions in this book pertain to ways of enhancing memory encoding. You do this by converting words into images, which come more naturally to us. If everything develops normally, we do not have to be taught how to see; we are born with that ability. But we have to learn—sometimes laboriously—how to read and write. It is the clarity and quality of the images formed that distinguish the superpower memory. Details about this will come in Chapter 3.

As mentioned earlier, emotion is probably the greatest stimulus to the creation of vivid imagery and the formation of memory. Suppose you outplayed your friends in the weekly golf game and couldn't wait to tell everybody. But when your ten-year-old daughter ran out to congratulate you coming off the 18th green, she was stung by a bee and needed CPR to, thankfully, fully recover from anaphylaxis. Which memory do you think is more likely to remain with you?

Time works to undo even the clearest memory. But there are compensations, as described by one of my favorite authors, George Orwell.

"In general, one's memories of any period must necessarily weaken as one moves away from it. One is constantly learning new facts and old ones have to drop out to make way for them. But it can also happen that one's memories grow sharper after a long lapse of time, because one is looking at the past with fresh eyes and can isolate and, as it were, notice facts which previously existed undifferentiated among a mass of others."

Finally, you should care about your memory because if you ultimately hone your memory skills and maintain them, you could protect yourself, to the extent that it's possible, from the ravages of Alzheimer's and other dementias. The key concepts in doing this are *develop* and *maintain*. This book will provide you the means of developing your memory. The maintenance portion of the process will depend upon applying my suggestions every day. The application needs to be not only effective, but also fun and challenging. My basic aim is to help you develop a memory skillset that engages your interest and arouses your enthusiasm.

One final point. At the present time there is no way of guaranteeing anyone that they may not eventually develop Alzheimer's. But I have never encountered in my many years of neurologic practice a patient with highly tuned memory skills who was suffering from Alzheimer's disease or any other degenerative brain disease. Put in another way, a poor memory doesn't necessarily imply dementia, but a well-functioning memory virtually eliminates a concurrent diagnosis of dementia. In other words, you have everything to gain by increasing your memory skills, and nothing to lose.

As a first step in developing a superior memory, let's explore how thinking about memory has developed over the centuries. When did people first care about memory? What methods did they employ to improve it? What were the theories underlying the importance of memory—and how relevant are these theories today?

In order to answer these questions, let's take a brief survey of suggestions that have been made by thinkers about memory over the past two thousand years.

CHAPTER II

EVOLUTION OF OUR UNDERSTANDING OF MEMORY

COLLAPSE OF A BANQUET HALL

Many current principles used to strengthen our memories can be traced back hundreds or even thousands of years. And although I don't think it's necessary to learn everything some of the wisest minds have written about memory over the years, some familiarity with their approaches and their concepts makes it easier today to understand and strengthen your memory.

To the ancients, memory was not just a means of preserving the past, but was considered a tool for what we would now call creative thinking. By constructing within the mind a structured and orderly memory, the ancients thought it possible to forge new thoughts and establish original and creative connections.

Commenting on the depth and power of his memory, Saint Augustine wrote: "The vast mansions of memory were treasured in innumerable images brought in from objects of every conceivable kind perceived by the senses. In these mansions are hidden away the modified images we produce when by our thinking we magnify or diminish or in any way alter the information our senses have reported. In the immense court of my memory sky and earth and sea are available to me, together with everything I have ever been able to perceive in them."

The Greeks were among the first to use special techniques to perfect the faculty of memory.

The key discovery can be traced to the collapse of a banquet hall. The poet Simonides (556–468 B.C.) was performing at a banquet and survived the collapse of the building (luckily he had been called outside a few moments before the collapse). Using his memory, Simonides was able to identify the dead based on his recall of the places where each of them had been sitting during the banquet. Traditionally, many have claimed that he was not only able to envision and name the positions of the attendees, but could identify what they were wearing and other indicators that differentiated one attendee from another. This remarkable performance suggested some of the principles of the art of memory. Here is Cicero's description of Simonides's insight:

"He inferred that persons desiring to train the faculty of memory must select places [the seats at the banquet hall] and form mental images of the things that they wish to remember [the identity of the banqueters]".

By restoring the images of the banqueters sitting at their places at the banquet table, Simonides made it possible to use the order of the places as a means of preserving the order of the individual banqueters. The key principle of Simonides's memory method was the formation of mental images coupled with their orderly arrangement.

THINKING IN PICTURES

Another Greek insight into memory can be dated to Aristotle (382–322 B.C.) in *De Anima*. In a profound insight, Aristotle asserted that perceptions are transferred to the mind (we would now say the brain) where they are elaborated into mental images. To Aristotle, the formation of mental images was similar to a tracing made on wax by a signet ring; whether a memory results depends on the conditions of the wax (the brain) and the signet ring (the stimulation). In this analogy the experiences conformed to the signet ring, and the wax to the brain.

"Some men in the presence of considerable stimulus have no memory owing to disease or age, just as if a seal were impressed on flowing water. With them the design makes no impression because they are worn down like old walls in buildings or because of a hardness of that which is to receive the impression. For this reason the very young and old will have poor memories; they are in a state of flux, the young because of their growth, the old because of their decay."

To Aristotle, images formed by imagination became the basis not only for memory, but for all thinking. Indeed, Aristotle believed that thinking cannot take place without pictures. "The thinking faculty thinks of its forms in mental pictures." Each memory becomes part of one or more networks associated with a particular past experience and each person possesses unique networks. For instance, while one person raised in a northern climate may associate the word "white" with snow, another person not accustomed to wintry weather would be more likely to associate it with "milk."

Over the next two thousand years, Aristotle's linking of thought with the formation of mental pictures gave rise to the development of various memory techniques, which we will discuss in chapter III. These are all based on the deliberate construction and manipulation of images. Most of these systems involve linking familiar places (loci, as they are referred to) with images of the things that the person is trying to commit to memory.

The underlying principle behind the use of mental imagery is a fundamental one. Pictures are easier to commit to memory than words. This is based on the fact that the brain wasn't designed for reading words; reading doesn't come naturally. We have to be taught how to read, while we require no instruction to form mental images of the objects and people around us. Historically, reading only gradually evolved in line with the emergence of Egyptian Hieroglyphics, The Phoenician Alphabet, and much later the Gutenberg Press.

As a result of the preeminence of images over words, the easiest way to remember anything, especially names, is to convert the name into a picture. But in order for this method to be effective, the images must be visualized with absolute clarity.

According to Cicero, "One must employ a large number of places which must be well lighted, clearly set out in order and at moderate intervals apart." The images associated with these places must be "active, sharply defined, unusual, and have the power of speedily encountering and penetrating the mind."

Over the centuries, various experts listed steps that could be taken to strengthen memory. In the thirteenth century one of the first, Boncompagno da Signa, an eloquent writer and expert on rhetoric (prose composition), provided probably the most eloquent definition of memory while he was teaching at the University of Bologna from the mid-1190s to the 1230s: "Memory is the glorious and wonderful gift of nature, by which we recall the past, comprehend the present, and contemplate the future through its similarities with the past." Even eight hundred years later, no one has defined memory with such elegance and concision.

Boncompagno da Signa's twelve characteristics through which memory is strengthened are: contemplation, study, debate, discussion, conversation, novelty, change, habit, rivalry, fear of criticism, desire for praise, and ambition for excellence.

Boncompagno da Signa even suggested five mental states that oppose the creation of a strong memory: useless worries, greedy ambitions, anxiety about those dear to us, drink other than in moderation, and excess eating.

After the twelfth century, the art of memory entered an eclipse, thanks to development of handwritten books produced by monks which largely displaced old memory systems. Why try to remember something when you can look it up in a book? But during the sixteenth century, the art of memory experienced a revival, thanks to Marsilio Ficino and Giovanni Pico della Mirandola. Both were members of a brilliant circle surrounding the Medici Court in

Florence. They were also cofounders of the movement known as Neoplatonism.

Neoplatonism was based on the writings known as the Corpus Hermeticum, attributed to the ancient Egyptian seer, Hermes Trismegistus. According to Marsilio Ficino, another Neoplatonist, Hermes Trismegistus was a charismatic figure thanks to his suggestion that the images formed by the imagination could transform the mind in the manner of a sculptor shaping stone. Thus, lofty and uplifting images could serve as a means of spiritualizing the mind and attaining "true wisdom." Based on this belief, the templates chosen by Ficinio and Mirandalo for restoring memorized material consisted of religious and mystical concepts.

Giordano Bruno, a renowned creator of memory systems, described in his two books *Shadows* and *Seals* various techniques for organizing the mind by establishing contact with cosmic powers. Intriguingly, the seals bear a striking resemblance to contemporary depictions of the brain's nerve cell networks.

MEMORY THEATRE OF GIULIO CAMILLO

By 1532, Giulio Camillo, a professor at Bologna, suggested a means for transforming the mind through a uniquely powerful memory system of his own creation. *The Memory Theatre of Giulio Camillo*, as it came to be known throughout sixteenth-century Europe, consisted of a wooden memory palace shaped in the form of a Roman amphitheater. To get an idea of how the theatre looked and how it worked, look at the short video on YouTube, "Conceptism and Giulio Camillo" by Mario Fallini (created by Flatus Vocis).

In Camillo's theatre, the spectator—representing the practitioner of the art of memory—stands on a stage facing the seats that are arranged as a seven-tiered structure with seven aisles extending from top to bottom. On each of the seven aisles are doors representing the seven planets. These doors are decorated with images of Cabalistic, Hermetic, and astral figures.

On the underside of each of the seats in the theatre are drawers containing cards that detail everything that was known at that time or even potentially knowable. Camillo wrote of his theatre that "by means of the doctrine of loci and images, we can hold in the mind and master human concepts and all things that are in the entire world."

In describing his memory theatre, Camillo compares the process of achieving wisdom via the cultivation of memory to the experience of being immersed in a dense forest. At first, the desire to see the whole extent of the forest is frustrated by the surrounding trees. But if a way can be found of ascending along the slope, it becomes possible to see a large part of the forest's form. When the top of the hill is reached, the entire forest can be seen. Camillo suggests that "the wood is our inferior world; the slope is the super celestial world."

Camillo's forest analogy makes clear that he intended his memory theatre to provide a means of ascending from a lower to a more spiritual realm. "In order to understand the things of the lower world, it is necessary to ascend to superior things from whence looking down from on high we may have a more certain knowledge of inferior things."

It's been suggested by classics scholars that Giulio Camillo's now-vanished wooden memory theatre represented a vitalized Renaissance version of the psyche, wherein humans participate in the life of the divinity via communication between the divine and the human. Camillo and his followers believed that humans possessed divine powers and are therefore capable, thanks to memory, of linking the everyday world with the divine world. As a consequence, new powers become available to the psyche.

"By means of the doctrine of loci and images, we can hold in the mind and master all human concepts and all things that are in the entire world," wrote Camillo.

In this process, images drawn from religion are imprinted on the mind with sufficient strength, that when a person bearing this imprint returns to the everyday world, the external appearances of that world became spiritually unified through the power of memory. As suggested by Francis Yates, perhaps the world's authority on the historical aspects of memory, "The images of Camillo's Theatre enabled the spectator by inspecting the images to read off at one glance the whole contents of the universe."

Notice that all of the methods for enhancing memories discussed so far emphasize the central importance of *concentration* and *repetition*. "If you pay attention (direct your mind), the judgement will better perceive things going through it (the mind)," according to an ancient fragment of the Dialexis written in 400 B.C. Later Saint Augustine referred to memory as *venter animi*, "a kind of stomach of the mind." He compared the establishment of memory by reading and re-reading a particular text to the action of a cow chewing a cud. In this pastoral comparison, readers "digested" their text by saying it aloud while meditating on it.

Rather than a "mindless" exercise, such focused concentration was considered a creative act leading to imagination, innovation, and invention. But before making an original creative contribution on one's own, it was necessary to master the works of earlier thinkers by committing their words to memory. "From antiquity, the arts of memory were conceived of as investigative tools for recollected reconstruction and selection, serving what we now call creative thinking," according to Mary Carruthers, an international authority on memory. "Memory craft was practiced as a tool of invention discovering and recombining things that one had previously learned."

THOMAS BRADWARDINE'S ADVICE

Here are specific instructions for forming such images by the fourteenth-century Archbishop of Canterbury, Thomas Bradwardine. The images of what one tends to learn should be: "wondrous and intense because such things are impressed in memory more deeply and are better retained. Such things are for the most part not moderate but extreme, are something greatly beautiful or ugly, joyous or sad, worthy of respect or derision, a thing of great dignity or vileness." In other words, in order for the images to be remembered they should be as dramatic and exaggerated as possible.

If we consider our own memory for a moment, Bradwardine's advice turns out to be true. We tend to remember things and events that arouse our emotions at the moment of memory formation. Thus we may not remember what we ate for lunch yesterday, but we will remember with great clarity a near accident we experienced while driving to the restaurant. "Extreme things excite the human senses and the human mind with greater force than do average things" wrote Jacobus Publicius in the *Art of Memory* published in 1475. "The reason for this is that things which are great, unbelievable, previously unseen, new, rare, unheard of, deplorable, exceptional, indecent, unique, or very beautiful, convey a great amount to our mind, memory and recollection."

Over the centuries, various objects served as place marks for memorization. "The domestic and familiar species of a Roman house—the type of architecture most commended for memory work by ancient writers—were replaced by divine structures derived from descriptions in the bible, such as the Ark, the Tabernacle, the Temple of The Heavenly City . . . the cosmos itself," according to Mary Carruthers, author of *The Medieval Craft of Memory*.

Metrodorus of Scepsis, a Greek scholar known for his works on rhetoric, relied on his knowledge of astrology to devise a memory system based on the twelve signs of the zodiac. Included with each of the signs of the zodiac were the decans: a subdivision of a sign divided into three equal parts of ten degrees each. Both the astrological signs and the three accompanying decans functioned as memory place marks. To increase the numbers of place marks, Metrodurus assigned ten background images for each decan. This resulted in a series of loci numbering from 100 to 360, which he could employ as a memory aid. (Three decans for each of the twelve signs of the zodiac multiplied by ten images assigned to each decan).

As Quintilian described it, Medrodorus "found 360 places at the twelve signs to which the sun moves." And since all of the loci were arranged in numerical order, Metrodorus could find any number of them by its number. By using this framework, Metrodorus could remember vast amounts of data and perform astounding feats of memory.

Perhaps at this point you are thinking *whoa! Enough already. That's pretty heavy stuff.* And it is. But it only hints at the power certain mystics and philosophers suggested memory could achieve. As you will see, these ancient and honored systems and beliefs have been transformed and still linger today as the basic tenets of the art of memory.

HERE IS HOW IT CAN WORK FOR YOU

Let me provide you with a personal experience of the art of memory as practiced by the medieval mnemonists. In chapter 3, I will set out an easy way to construct your very own personal memory system. Here is a brief foretaste of that more detailed construction:

First, let's try committing to memory five nouns chosen at random. After forming your image of the nouns, place them on five familiar places chosen from, say, the furniture in one of the rooms in your home. See the objects in your mind's eye occupying those familiar places. If you can do this successfully, it should be easy to mentally walk from one location to the other and see the five objects in the location (or loci, to use the original Latin term) where you mentally placed them.

When performing this preliminary exercise, I selected the following words completely at random: *seed, enemy, train, juggler,* and *watch.* For the places where I would store my five images, I selected five memory loci from my office: my desk, the waiting room, the hall leading from my waiting room to the outside door, the entrance to the building, and finally the museum directly across the street from my office.

In order to remember the words, I chose dramatic striking images involving my office in order to increase the likelihood that I would remember them. I imagined a huge *seed* sitting on top of my desk and extending up to the ceiling. As I walked from there to my waiting room I encountered a man with a knife (*enemy*). I slipped past the knife-wielding man and ducked into the hall where I discovered my exit blocked by a red *train*. I skirted around the train and opened the door to the outside where I encountered at the entrance to the building a *juggler* in a purple outfit juggling oranges. When I looked across the street to the museum I discovered that the building had been transformed into a huge *watch*.

Going forward, when I remember those five words all I have to do is mentally retrace my steps and encounter the five strange and improbable images I created. Now take the five words you selected and link them to five places you choose as your memory loci; try to come up with equally dramatic and unusual associations. The images you form as a help to remembering your words can come from any source familiar to you, based on your background and experience. More on how to do this will come in the next chapter. But first, a pivotal question: How and when did memory evolve from an art into a science?

A SHY BEARDED INTROVERT

Scientific experiments on memory began with a bearded, shy itinerant professor, Hermann Ebbinghaus. His goal was to discover how memories are formed and how they fade. In contrast to earlier thinkers specializing in memory, Ebbinghaus was not a philosopher, but what we would now call an experimental psychologist. His goal was to move beyond speculation and put memory on a more firm scientific foundation.

At the time of Ebbinghaus's experiments—the mid-nineteenth century—a lot of emphasis was placed on rote memory: learning by constant repetition. The memory aspirant who wanted to learn the multiplication tables for instance, would repeat them over and over until they could be recited on command. The same repetitive process would be expanded to learning such facts as the geographical locations of world nations and their capitals. This emphasis on rote learning continued until well into the twentieth century. Anyone who was in primary school in the 1950s and 1960s can attest to the fact that rote memorization was still enforced and heavily relied upon.

Today we tend to de-emphasize rote learning, "poll parroting" as it's often referred to. As a result, it's not unusual to encounter grade school students who can't name any state capital other than their own—and sometimes not even that one. But the nineteenth century emphasis on rote memory made it natural for Ebbinghaus to carry out his experiments employing rote memory.

Stimulated by Lewis Carroll's *Through the Looking Glass* (1872), with its inclusion of verses of rhyming nonsense words, Ebbinghaus decided to use three letter-word combinations which had no meaning or associations. They were composed of Consonant-Vowel-Consonant combinations, what Ebbinghaus referred to as CVC trigrams made up of two consonants with a vowel between them. The two consonants must always be different (TAC was ok, but TAT was not) and the trigrams could not express any meaning (TIP or COT were not admissible) but only had the form of a word. After working for months applying these two rules, Ebbinghaus created twenty-three hundred CVC combinations. From this stockpile, Ebbinghaus wrote varying numbers of the trigrams in his notebook and, to the regular beat of a metronome, he did his best to recall them.

By this means of this time-consuming, ponderous, and, for anybody but Ebbinghaus, utterly boring process, he discovered two key facts about memory. First was the *forgetting curve*: the loss of information that occurs after the establishment of a memory. The sharpest decline occurs in the first twenty minutes with additional decrease within the first hour. It then levels off after about a day. If you can still remember a fact or an event after a few months then the odds are quite good that you will remember it indefinitely.

The *learning curve* is a measure of how fast one can learn new information (in Ebbinghaus's case his recall of 2300-letter combinations). This too followed a pattern. Ebbinghaus read a list of twenty nonsense words out loud to himself and put away the list for a period of time then tried to recall as many words from the list as possible. After checking for the words missed, he repeated the process until he could remember all twenty nonsense words from the list.

Ebbinghaus's experiment involved the key processes underlying memory formation: learn, delay, test, relearn, delay, retest. Retention drops quickly over the first few days and then tapers off more slowly. If the delay between learning, testing, and relearning is short, nearly 100 percent of the information can be retained. But as the delay increases to four days, the retention rate falls to 25 percent. The longer and more intensely he practiced, the better his later recall.

Especially intriguing was Ebbinghaus's discovery that he could often correctly select a previously seen nonsense word when it was inserted into a list of other nonsense words he hadn't previously encountered. He could do this even when he could not spontaneously remember the word if asked to do so. This surprising superiority of *identification memory* over *free recall memory* allows us to recognize the correct answer from a list of possible choices even though we are not able to spontaneously come up with the correct answer on request. As mentioned earlier, this is the basis of the multiple choice method of testing that is now so popular on standardized testing.

Ebbinghaus's unconscious memory for words that he had seen but couldn't consciously recall was based on a process psychologists would later refer to as *priming*. Parts of a word (WA_, BA_) are shown to a volunteer who is asked to fill in the blanks with the first word that comes to mind. Most people randomly choose common words (Water, Barn). But if a person has previously read a list of words starting with the letters WA and BA, they are much more likely to fill in the blanks with letters taken from words included on that list (i.e., Wasp, Bank, if those words had been on the list). This occurs even though the person has no conscious memory of having seen the words on the list. Psychologists refer to this unconscious memory as *tacit knowledge*, which we all have. As another way of putting it, we all know more than we can say.

Both Ebbinghaus's experience and the priming experiments point to an important distinction between remembering and knowing.

Think back to the last time you met someone who seemed very familiar to you, but you could not precisely identify them or remember the context in which you first met them. In order to jog your memory, you engaged the person in conversation. Your hope was that during that conversation a specific item of information would arise that could serve as a clue for you to remember more about that vaguely familiar person. Here is what was happening: Your initial feeling of familiarity about this person was based on an inaccessible memory you retained from sometime in the past. Yet this feeling of "knowing" wasn't sufficient. Additional clues were required until knowing flowers into remembering.

Knowing but not quite remembering forms the basis for the "tip of the tongue" experience. We have all experienced it at one time or another: We can almost but not quite come up with a specific memory, usually the name of a person or object. This difficulty is the result of imperfect storage. You know the answer, it's just at the point of expression, but you can't quite access it. But by gathering more information and linking this information with other memories that we established at an earlier time, we are able to convert a frustratingly vague sense of knowing into a positive identification. Ebbinghaus described this as *savings*, which is the amount of information retained in our subconscious even though this information cannot be consciously accessed.

As an example, Ebbinghaus would memorize a list of items until he could recall them perfectly and then he would not access the list at all until he could no longer recall any of the items. He would then relearn the list and would compare the time required for relearning the list. He found that the second memorization was faster and the difference between the two represented the "savings." The former occurs "with apparent spontaneity without any act of the will"; the latter brings memory into consciousness by "an exertion of the will." *Savings* represents this difference between voluntary and involuntary memory.

As Ebbinghaus discovered, memorizing by rote (repeating something in the same order over and over) is very limited, difficult, and prone to failure. This holds true not only for nonsense syllables, but for real words conveying meaningful information. Think back to school when you were learning to recite the lines of a poem. If you forgot one or more lines, you could retrieve them only by starting the lines (or sometimes the whole poem) over again and hoping that the words would spontaneously come to mind. You were hoping that memory for the words and lines were lurking somewhere within your brain in an unconscious non-retrievable state and would spontaneously spring to mind. Frequent failures in this suggest that rote memory is very limited and confined because it is based on a single-access approach to the memorized material.

As is evident from Ebbinghaus's groundbreaking experiments, the brain isn't fashioned for rote memory—that's the principal reason we find brute memorization so difficult. Rather, the brain operates using *associative memory*, where things are linked with one another via associations. This is much more adaptable and creative than rote memory. For example, suppose that during our conversation you suddenly can't come up with a specific word to describe what you are trying to say. In most cases you can't remember the word by simply trying harder. But if you think of a word frequently associated with it, you may come up with that elusive word.

FDR AND THE BRAIN

While passive rehearsal works over the short term, it's not effective in transferring from short-term memory to long-term memory. For instance, you read a telephone number taken from your television screen, say, an advertisement for a watch or piece of jewelry. You keep repeating it to yourself while you head upstairs to retrieve your cellphone. In most instances, this internal repetition works just fine. You remember the number and successfully dial it on your cellphone. It's not likely that you will remember the number for more than a few minutes afterwards. But that's ok. You didn't intend to store the number for the long term.

Short-term memory provides the holding place needed for keeping information in mind for immediate use. It is important to distinguish this memory that forms and disappears (transient) from memory that you want to have available over the long term. But before we move into the different forms of long-term memory, let's look at what happens when a memory is formed.

As a first step, the memory to be recalled—say, one of Franklin Delano Roosevelt (FDR), the president during World War II—is encoded when first learned within the hippocampus, the seahorse-shaped structure in the temporal lobe that connects via an arching circuit, the fornix (Latin for arch) to two other structures, the mammillary bodies, and the dorsal thalamus. An additional structure—the basal forebrain, located, as the name implies, at the base of the anterior portion of the brain—is also linked to the fornix. And finally, slightly behind the hippocampus is the amygdala, the almond-shaped structure that endows memory with emotion. We don't just remember things; we also experience emotions associated with our memories. It's emotions that anchor our memories. Whenever you exercise your memory via memory challenges, you strengthen the linkages of these structures in this circuit and the stronger your memory becomes.

After the Franklin Delano Roosevelt information was stored within these memory-encoding structures, that information was next widely distributed to the rest of the brain via association fibers.

Neural networks is the term used by neuroscientists when describing the linkage of association fibers. As the stimulus enters our brain, it creates an electric and then chemical event (electro-chemical signal) that races along the length of a neuronal axon (an extension of the neuron) like a Formula One car. The impulse is then transferred to the next neuron by means of the release of chemicals from the first neuron (the presynaptic neuron) into a tiny cleft (the synaptic gap), where it is ferried across by various transporters to the second neuron (the post-synaptic neuron), which is then activated. The process repeats itself again and again from neuron 3 to the nth neuron depending on how many neurons were originally required to form the memory.

This lattice-shaped network that forms in a particular brain differs from the networks of brains of every other person in the world. No two people experience the world in identical ways and therefore individual memories and the neuronal networks comprising them will always differ. Think of it as delicate and intricate gold sculpting—only the fundamental substance in question isn't permanent like gold, but more evanescent like a flash of lightning. These neuronal networks that form the foundations of our memories change over our lifetime. This explains why memories change in subtle ways even from day to day. In fact, the neural network is altered electro-chemically whenever a memory is brought to mind and then passes away as we move on to other matters.

In general, long-term memory storage occurs in select areas of the cerebral cortex. Your recall of Franklin Delano Roosevelt's name involves language and vocabulary stored and processed chiefly in the association areas in the left temporal lobe. Your recognition on recordings of Roosevelt's stirring voice urging his fellow Americans to heroic efforts is stored within the right temporal lobe. Think of your experience of Franklin Delano Roosevelt as involving many distinct brain modules working together to produce the totality of all the things that you have learned and can state (i.e., declare) about Roosevelt. This type of memory, *declarative memory*, is based on the simple fact that the knowledge can be declared. If someone asks you a question about Roosevelt, you are able to answer (declare) it based on your long-term stored information.

Under normal circumstances, the memory components are parceled out to separate areas of the cortex towards the top of the brain (sight to the visual cortex, sounds to the auditory cortex, etc). These separate components don't act alone, but communicate directly with each other via a network of interconnecting centers. Especially important are the frontal lobes, which are part of the anterior and midline structures of the brain. More about them in Chapter III.

Damage or interference with any of the structures in the memory circuit results in various forms of memory disorder. For instance, if the hippocampus is involved, then a memory cannot initially be encoded. We will take up examples of the effects of damage to the other memory structures in chapter V.

Based on the findings of modern neuroscience, we now know that all of the things that we commonly associate with memory—all the names, facts, and figures we have learned through life—involve specific brain structures linked together in the circuit, which I have just described.

ALBERT EINSTEIN DRINKING A CUP OF COFFEE

Many things have changed since the bizarre experiments of Ebbinghaus. We now know that memory depends on associations rather than single words. Each word has to be put in context and associated with other words or phrases in order to form a memory for later retrieval. So your best chance of remembering is to enlist the brain's powers of association.

Three principles underlie the formation, retention, and recall of a memory: 1) Multi-coding 2) Organization 3) Association. As you will see, all three principles reinforce one another.

1) Let's take up multi-coding first. Picture yourself drinking coffee. Not only can you imagine yourself doing that, but you can also imagine smelling its delightful aroma. In your imagination you can taste it, savor it as it flows over your taste buds. The coffee experience is both verbal (you can describe and name the coffee), as well as imaginative (tasting, smelling, etc.). Most proper names are like this. "Chair" and "notebook" can be described and imagined in different ways (the comfort of the chair, the softness or hardness of the notebook, etc.). The more senses that can be recruited—*multi-modal coding*—the more likely you will be able to form a long-lasting memory. The reason? More areas of the brain are involved. When you say the word that you are trying to memorize, you are calling into play the language centers in the left hemisphere. When you touch, hear, smell, taste the object—or just imagining what the object would feel like, sound like, smell like, and taste like—it recruits the brain areas mediating these sensations. So if you were imagining a comfy cup of coffee moments ago, you were emphasizing the same brain areas, as you would when you actually drink it.

Now consider a word such as *epistemology*, which can only be expressed verbally. No image or taste or smell accompanies saying or reading epistemology. That's what makes such terms difficult to commit to memory. In order to remember them, what's needed is to create an imaginative link involving any of our senses.

The art of memory requires mentally transforming abstract concepts into images, and bringing these images alive, as if they were real physical objects. If you try to remember to take your car keys, you must make yourself see them in your mind's eye, feel their weight, experience the sensation of putting the key in the ignition and starting the car—all in your imagination. "The ability to modify imaginary objects goes hand in hand with the capacity to represent them," according to philosopher Paolo Fabiani. As previously mentioned, images formed in the imagination are most effective as memory prompts when they are rendered whimsical, inappropriate, even bawdily outrageous compared to the real objects that inspired them.

So how might you go about representing a word like *epistemology*? Think back to that coffee mentioned a moment ago. Picture in your mind Albert Einstein holding a cup of coffee in his hand while reading a book titled *Epistemology*. This image of the world's most famous scientist intently reading a book dedicated to how justified belief can be distinguished from mere opinion (a workable definition of epistemology) will provide you with an accessible memory for the meaning of epistemology. The smell increases alertness and concentration—two qualities Einstein would require for delving into such a heady subject as epistemology. The guiding principle: Our memory can be enhanced by using our imagination to link the thing we are attempting to remember to images that highlight it.

2) Organization is important because a successful memory performance can't take place without it. Our brains are designed to work with *meaning*. If meaning isn't obvious, we create it (as good an explanation as any to account for the existence of conspiracy theories). A list of a dozen unrelated objects is difficult to remember for the long term without some form of organization. The easiest way to organize unrelated information is to associate the things you are trying to remember with something you already know. Organization involves creating a framework for transforming random information into something meaningful and therefore easier to memorize.

As an example, I parked my car yesterday in parking space 351 in a seven-story garage. The penalty for forgetting that number could turn into an exhausting daytime nightmare. I shudder when I picture myself walking along what seems an infinite line of cars trying to identify my own. So how to guarantee that I will be able to find it? By using the sounds-like system based on rhymes. The number "three" rhymes with "tree," "five" rhymes with "hive," and "one" rhymes with "sun." I pictured a tree in full bloom with beehives so numerous and ponderous that they weigh down all the branches. This scene took place under a blazing sun. When I returned to the garage I had no trouble converting the images back to numbers. More on this in chapter III.

Finally, 3) association is basically a blend of multi-coding and organization. Simply thinking about how two or more things can be associated requires you to concentrate and focus—two brain activities which on their own lead to enhanced memory.

MIND MAPPING AND THOUGHT TRACKING

Another method of association involves thought tracking. When we are conscious and thinking over even a short span of time, we are likely using our working memory, defined as a mental workspace where short-term memory can be manipulated. Here is a way of experiencing. You only need a digital watch with a timed alarm function; your omnipresent cell phone can serve just as well. Have a friend set the alarm to go off at a time unknown to you several hours in the future. The only requirement is that you be awake when the alarm goes off and probably not at work, or otherwise responding to professional or domestic demands.

At the instant that the alarm sounds, take careful notice of your thoughts. Try to recall the thought immediately preceding the alarm and then the thought before that one, and then the thought leading up to that thought and so on. Track the linkages as far back as you can. Don't guess and don't make anything up. All the thoughts that you recover are taken from discreet elements in your working memory. At first, you probably won't be able to trace the content of your consciousness more than a few steps. But with practice you will be able to track back over a dozen or more associations. Your aim in this exercise is to increase your critical and observational powers, enhance your pattern recognition, bolster your synthetic and creative skills and, finally, improve your ability to communicate in ways that will be comprehensible to others. This chain of events can be started anywhere and tracked either backwards or forwards. You can begin anywhere in the chain of associations created by your brain.

The nineteenth century psychologist William James described this process, "start from any idea whatever, and the entire range of your ideas is potentially at your disposal. . . . The entire potential content of one's consciousness is accessible from any one of its points."

SUNGLASSES AND LIPSTICK

Another method of thought tracking involves writing down all the associations that occur to you in response to seemingly unrelated events, topics, or objects. As an example, here is a paragraph based on the linkages produced by my thoughts over a fifteen-minute period after paging through a magazine. I selected two pictures from the magazine at random taken from advertisements of seemingly unrelated products. One was a picture of sunglasses, the other a picture of lipstick. Here are the associative links that I came up with:

Sunglasses were developed to shield the eyes from the harmful effects of the sun, while lipstick protects the lips from dryness, a byproduct of the exposure to the sun. Both products resulted from advances in synthetic chemistry. But synthetic chemistry also created polystyrenes and aerosols, the chemicals found in deodorants and shaving crèmes. Increase in the use of these products is depleting the ozone layer and, as a byproduct, increasing the instances of cataract disease and lip cancer secondary to sun exposure. But polystyrenes also made possible the development of the photographic film to capture the images of the sunglasses and lipstick that I looked at in the magazine. Sunglasses lend to their wearer an element of interest, mystery, or glamour (Anna Wintour), which increases the appeal of sunglasses and, as an unintentional byproduct, induced increasing numbers of people to wear the glasses and thus cut down their chances of contracting sun-related cancers. But sunglasses are also sinister (Mafia dons are unvaryingly depicted in sunglasses; Idi Amin wore reflective sunglasses, so that his victims could only see their terrified expressions reflected back at them). Amin and the mafia are associated with death and their dark glasses suggest the inhabitants of Hades. Used in the singular, a "shade" is a visor for shielding the eyes from strong light, and, hence, a forerunner of "shades," the colloquial term for sunglasses. But a shade is also a scientific apparatus or shutter for intercepting light passing through the camera that enabled the photographer to take the two pictures of lipstick and the sunglasses that I'm looking at.

Pick any two items at random and try this associative method yourself. No two individuals will form the same patterns because no two individuals possess identical brains or have undergone identical experiences. My own associations to the sunglasses-lipstick example were no doubt heavily influenced by my scientific and medical education. If your background is in the humanities, then polystyrenes, ozone layers, and cancers aren't likely to be among the words that first spring to mind when you associate sunglasses with lipstick. More likely, your patterns will reflect your own personal experience gained from literature or any of the other humanities. Go ahead and do the sunglasses-lipstick exercise using your own associations.

A more formalized method of association is suggested by memory expert Tony Buzan.

Start by drawing a circle on a blank piece of paper. Put a word in the empty circle. Now surround that circle with additional circles in which you have written words and images that are associated in your mind with that word. Think of some of the concepts that may be associated with it. Each of these words and images are part of your brain's association network or "memory bank" concerning the word. By adding as many attributes that you can come up with, you will soon have constructed what Tony Buzan refers to as a *mind map*. This mind map increases the likelihood that you will successfully come up with the elusive word.

In one experiment on mind mapping, several participants were given the word sequence dog-bone-m. The second group was given the sequence gambler-bone-m. The subjects were then tested as to which of the two groups would come up with the response "meat" more quickly. The people in the first group dog-bone-m were faster, because the proceeding word "dog" activated the memory link "dog-bone-meat". As Tony Buzan suggested in his many books on the subject of mind mapping, "Memory works by an activation process which spreads from word to associated word via the links in the mind map."

Mind maps take advantage of the associative powers described by Aristotle. He was one of the first to espouse the principle that memory is based on the formation of linkages or associations. For instance, if I ask you to respond with the first word that comes to mind when I say "high," you are likely to respond with "low." In most people's minds, these words are associated ("hot" and "cold"; "wet" and "dry" are other examples). Aristotle suggested that associations are based on three principles. First, the more we experience two things together the more strongly we associate them. Second, events experienced at the same time or place tend to be associated. Third, if two things are similar the thought of one will likely prompt the thought of the other.

Over the ensuing centuries Aristotle's ideas have been supported by psychologists and neuroscientists. And while Aristotle had no way of knowing this during his lifetime, the associations once established involved the maintenance of networks of brain cells (neurons) that functionally linked together whenever we learn new information. Later, whenever we recollect a specific memory we activate the brain cell network that was formed at the moment when we first established that memory. The more frequent we access that memory, the stronger these linkages become. Some memories are so well established (such as our name, our addresses, our siblings, etc.) that they remain a permanent part of our identity. This is what we mean when we say that memory forms the foundation for identity.

When applying a mind map to recover a memory (a name or word, etc.), the central image will be blank and the associated nodes will consist of whatever springs to your mind in regard to that name or word. Where and when did you meet that person; who were some of the people familiar to both you and the person whose name resists your recall? You will find that without writing many things down or creating a lot of nodes, the name will come to you from your "unconscious memory."

Here is a short example of how you can recover momentarily forgotten words. Draw a node and leave it blank. I'm going to provide a word for that blank. But first, I'll suggest some nodes to you that should lead up to you guessing the word. The first node is *medical term.* The second node is *infectious disease*, which branches off directly from *medical term.* For the third node, *historical term*, make it a separate node branching off *from the blank node.* Next draw a fourth node, *plagues,* which splits off from both *medical term* and *historical term*

Finally, draw a last node, *compulsory separation of sick from well,* emanating from both *historical term* and *plagues.* Stare at diagram 1 illustrating the resulting pattern. Keep your eyes fixed on it for ten or fifteen seconds and then look way. Have you thought of the missing word now?

The word is all too familiar to us from our Covid-19 national experience. In this exercise, since I was the only person who knew the word, I functioned as a facilitator of your unconscious memory and, by means of mind mapping, helped you come up with the word ("quarantine") based on its associated nodal links.

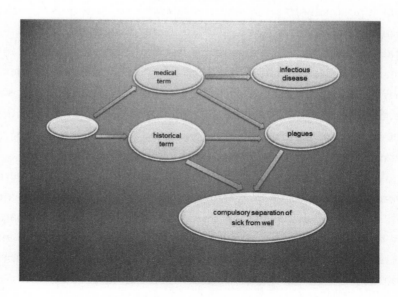

Since mind maps can play a vital part in strengthening memory, here is a practical example of the mind map I used to compose a short published essay on the subject of this book, "Memories" or, to use its published title, the nature of "Real Memories." See the diagram of the mind map which directed my thought processes in creating the memory essay by associating all the concepts that occurred to me at the time:

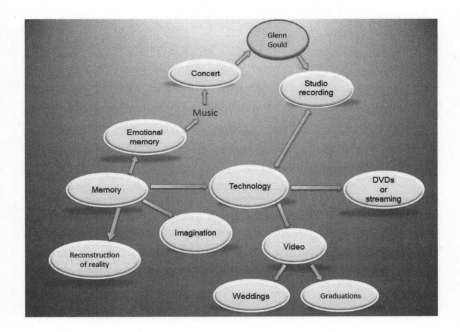

Notice how the nodal topics are introduced in spontaneous nonhierarchical fashion. Now here is the essay suggested by the mind map:

At a recent wedding reception I observed that the bandstand was occupied not by four or five musicians but by a disc jockey employing a carefully coordinated ensemble of speakers, microphone, vinyl records, CDs, and tape decks.

"How could they do this?" I asked myself, annoyed since I had been looking forward to live music. But an hour later, the whole wedding party, myself included, was on the dance floor frenetically responding to

the electronic music of a rock group that only someone with a major interest in a Fortune 500 company could have afforded to hire for his or her daughter's wedding reception.

The next morning, after making rounds on my patients, I decided to pause for a moment in the chapel of Georgetown University Hospital. I entered to the sounds of monks chanting. But there were no monks. A stereo system was making it all possible. The monks' presence, however, was palpably real. Indeed, what would have been added to the moment, I asked myself, had the monks actually been there?

Which is more real: the live performance (by rock groups or monks, take your choice) or a recording? Audio and video technology is making that question very difficult to answer. Take for instance, the enthusiasm for hiring videographers to record weddings, graduations, and anniversaries. With the passing of the years it's likely that these recordings will become more real to the participants than their own fading memories of the events they participated in.

As this process of forgetting progresses, a point eventually will be reached when the marriage or graduation will be remembered almost entirely from the video—and better than the event itself.

Think back to the last time you attended a concert. Compare the experience to hearing that same music on a good stereo system. Which is more real? Glen Gould in his later years gave up live performances to concentrate entirely on recordings. To him these were more real, closer to the reality that the composer and Gould, as performer, were attempting to get across.

Technological productions, audio and video specifically, are approaching a degree of precision and verisimilitude greater than the live events they record. If I closed my eyes in the chapel I could be convinced that somewhere in the shadows monks were chanting in perfect unison (much better, no doubt, than most ensembles of monks that could have been assembled for the occasion). If the measure of reality is "felt" reality, then this recorded performance by anonymous monks was accomplishing its purpose: I felt rested, at peace, in a mood of somber but pleasant introspection.

Reality, it seems, is sufficiently fluid to accommodate whatever trans-formations are required by us to make the past conform to our needs in the present. For instance, all the Christmas presents that I can recall from my childhood I remember not according to the order in which I received them, but according to their importance in my present emotional balance. I still recall my dog and my bike, but other gifts of little emotional appeal to me now are lost in the mists of forgetfulness.

Such selective reconstructions of the past are the most threatened by the new technology. What really happened so many Christmases ago will never be what was captured on the video. Instead, that memory will remain the much more delicate and evanescent interchange between my experiences then and the modifications of those experiences modeled within the neuronal networks of my brain. The past, like the present and the future, is fluid and dynamic, changing according to our emotions and interests at the moment.

SITUATIONAL AWARENESS

Situational awareness exercises are used by the Navy Seals and other branches of the military. On request a seal must be able to describe the location of the doors and windows of the room in which he is sitting, along with other details that would be helpful to remember in order to make a quick escape in the event of an enemy attack.

To get a feel for this, the next time you are in a restaurant, close your eyes for a few seconds and mentally picture the arrangement of the people sitting around you at the nearby tables. If you are like most people, you probably won't do very well with this memory exercise the first time you try it. You have to train yourself to increase your attention to what's going on around you. The goal is to employ your attentional focus in a manner of a search light scanning the night sky. The more you practice this exercise, the greater the breadth and depth of your memory. You will see more and remember more because at a given moment, your memory is encompassing larger swathes of your immediate surroundings.

"If you are more aware of what is happening around you, you are not only experienced more in the present moment, you also have more memory content," according to Mark Wittmann at the Institute For Frontier Areas of Psychology and Mental Health in Freiburg, Germany.

One step beyond situational awareness exercises directed *outward* are situational exercises directed *inward*. Situational exercises involving self-exploration are used in creative writing seminars. After encountering unfamiliar people in a social setting, the aspiring novelist is asked to incorporate them into the plot of a novel or short story. A similar method is used in the training of psychoanalysts. It is referred to as self-analysis. The first patient was none other than Sigmund Freud. No one analyzed the Meister. He analyzed himself via the associations suggested by situational exercise tracking of his thoughts.

In the next chapter, we are going to parse the umbrella term *memory* into its component parts.

CHAPTER III

DIFFERENT TYPES OF MEMORY

DANCERS SWAYING TO THE MUSIC OF TIME

Our memories—the facts we know and the events we can recall from our past—form the basis of our identity. When we lose our memories, aspects of our identities change or totally disappear. "A person's memory is everything, really. Memory is identity. It's you," writes novelist Stephen King in *Duma Key*.

Since our identity is rooted in our experiences, the more experiences we can remember the richer our sense of ourselves. That explains why loss of memory is the most agonizing aspect of Alzheimer's disease. The progressive memory failure that occurs in this dreaded disease destroys not only memory, but a person's identity.

But when you come to think of it, what makes you *you*? Philosophers have debated this conundrum for centuries. In the seventeenth century, John Locke defined identity in terms of memory. As he pointed out, recollection provides a thread connecting our past with our present. Most of us register our day-to-existence, our sense of self, from how we are today which, in most instances, is pretty much indistinguishable from how we experienced ourselves yesterday. But if we go back far enough, this continuity of self-experience becomes more tenuous or lost all together.

In the 1970s, philosopher Derek Parfit argued that the links connecting our past to our present form memories like the links in a chain. In a phrase we remain "ourselves" because of the chain of experiences which becomes even more tightly bound by introspection and self-knowledge with our *memory* consisting of all of the links connecting our present to our past.

The gist, then, is that memories are the basis for personal identity. In a later chapter, we will take up the price that must be paid in regard to our sense of selfhood if these memories either disappear or undergo drastic alteration.

What is your earliest memory? Think about it. How far back in your past can you go and retrieve a clear memory? Most people can't remember anything prior to age two or three. In explanation, Freud popularized the concept that people can't remember anything before that as a result of repression (the unconscious preventing memory from entering conscious recollection). But today, neuroscience provides us with a more satisfactory explanation. We don't remember anything earlier than age two to age three because the brain structures responsible for episodic memory (things personally experienced) aren't sufficiently developed. Basically, we cannot remember anything prior to that early stage in our brain's development because we possess no sense of self.

Think of memory and identity like two dancers swaying to the music of time. Prior to age two, infants have no real identity and can't recognize themselves in a mirror. Within a year after that, they will look at the mirror and reach up to touch a spot of rouge dubbed on their face moments earlier by an experimental psychologist. At that point they're able to recognize that they are seeing in the mirror none other than themselves.

By five years, all the memory systems are online and functioning. Coincident with the maturing of the pre-frontal cortex and hippocampus—they are both late maturers—episodic memory (for events that have been personally experienced) and semantic memory (general knowledge) emerge along with language. It is language that super charges the infant's sense of identity. As remembered experiences slowly accumulate in the infant's brain with the help of language, the components interweave into a fine tapestry composed of the skeins of identity and memory.

The sequence is → words → concepts → identity → memories.

Both episodic memory and semantic memory are based in the cortex, which is involved in a two-way communication with the hippocampus. Over our lifetime, the hippocampus along with the specialized memory centers in the cortex perform a kind of call and response routine. If we are trying to remember our college graduation, for instance, nerve impulses proceed from the hippocampus to the cerebral centers in the cortex where the memory is stored (the separate sights, sounds, discussions, etc). And then the separate components are wired back from the hippocampus, which recreates the graduation experience in the form of an episodic memory.

Over time, identity, language, and memory create links between the present, the past, and the future. As a sign of life's ironies and cruelties, these same communication channels unravel towards the end of life in the same sequence as they developed during the first five years. The initial indication of Alzheimer's disease is usually a language problem, a loss for words (aphasia) and concepts, followed by the loss of identification for others, and finally, in the last sad stages of this dreadful illness, the failure to identify one's self.

"THE MAGIC NUMBER SEVEN"

There are different types of memories just as there are different types of dogs and cars. A poodle differs from a Great Dane, just as a Rolls Royce differs from a Prius. Each shares similarities to the other, but each also displays significant differences.

Memory can divide it into *transient* and *long-term*. If you remember for only a few seconds something that you have just seen, heard, tasted, felt, etc., we speak of transient sensory memory. Unless you make a deliberate attempt to retain the fleeting impression thus created, transient memory, as the name implies, rapidly disappears.

Memory for the taste of chocolate ice cream comes easily, but it's difficult to maintain it in memory for more than a few seconds.

To summarize, short-term memories are those that we hold in our mind's eye through active rehearsal. If I tell you a telephone number to call on your cell phone which is in the next room, you will probably store it in your mind through active rehearsal. In order to retrieve the number as you move into the next room, you will find yourself repeating it over and over to yourself. If anything interrupts you on your journey—maybe even the ringing of the telephone you are coming to retrieve—the number is likely to be either mistakenly recalled with the numbers out of order or forgotten completely. When you were told the number, it took the form of a short-term *declarative memory:* a clear conscious piece of information conveyed to you.

In this instance, the telephone number was only intended to be remembered for a short period of time—until you get to the phone and dial it. But if you want to include the number within the list of people you wish to stay in touch with, that episodic memory will have to be transferred into semantic memory: general knowledge learned through repetition. If you dial the number frequently enough it will be transferred into semantic memory. This transition may happen slowly—requiring multiple repetitions—unless the number is very important to you and/or is accompanied by some emotion.

34937052722275004546802087134565537006781921652344 5680
7561450359492340096067659 0
87

Look at the above string of digits for one minute. Turn away and write down as many as you can remember, starting from the left. How many were you able to come up with? According to the Montreal Cognitive Assessment (MOCA), the Gold Standard among neuropsychological screening tests, your performance was acceptable if you remembered 5 or more. But what is the upper limit of numbers that can be remembered by a group of people randomly chosen?

In the 1940s, a Harvard psychologist named George Miller discovered the numerical limit (seven) while designing a jamming signal for the armed forces aimed at disrupting German radio communications. In the process, he measured how people judge the magnitude of various physical stimuli. How loud is it? How bright? Miller found that people's ability to make judgements across a range of stimuli is limited to about seven alternative states. When Miller measured the capacity of people's short-term memory for digits, he discovered that it was also seven. He read a string of numbers and asked his subjects to repeat the numbers. He found that most people could repeat strings of five to nine numbers with the rare exception capable of learning strings of ten or more digits (in a few pages, I'll show you how to repeat more than ten digits on your first try!).

Noting that the number seven appeared both in the sensory magnitude rating (how loud?) and the digit span, Miller wrote in 1956 one of the most whimsically titled and famous papers in psychology, "The Magical Number Seven, Plus or Minus Two." (Interestingly, Miller was not the first to observe this. The nineteenth-century philosopher Sir William Hamilton pointed out that most people experience difficulty with numbers containing more than seven digits: "If you throw a handful of marbles on the floor, you will find it difficult to view at once more than six or seven without confusion.")

Later, research by Miller's students and colleagues showed that the "magic number seven" applied to digits, words, pictures, and even complex ideas. The take-home message from Miller's work? The brain works within certain limitations, and these limitations hold true over a wide range of human endeavors.

CHUNKING

But those limitations can be overcome. For example, let me show you the most basic memory organizing principles of all—guaranteed in one step to increase your immediate or short-term (these are alternative terms) memory. Take another look at the digits you attempted to memorize a moment ago. Think of the strings as composed of phone numbers.

349-370-5272,
227-500-4546,
802-087-1345,
655-370-0678,
192-165-2344,
568-075-6145,
035-949-2340,
096-067-6590,
87

It should only take a few minutes for you to get the first three phone numbers. But even if you can only remember two of them, you have already achieved a memorization of twenty numbers, far higher than the five numbers on the standard neuropsychological test (MOCA) or the number span of seven discovered by psychologist George Miller. The organizing principle at work here is called *chunking:* converting random numbers into a memorable string, in this case by learning them as telephone numbers. Notice that you didn't have to wrack your brain in order to come up with a way of imposing meaning on a meaningless sequence. Rather, when I suggested telephone numbers everything fell into place. Other methods of chunking will require you to think more intensely and consider the numbers with a greater *depth of processing:* the level of mental activity devoted to organizing new information. If not telephone numbers, then what? Let's explore that question.

You will find that the more effort you put into processing—searching for novel ways of representing the numbers, such as the telephone number model—the more likely you are to remember it. I suspect—but have no proof to support my suspicion—that Ebbinghaus managed his achingly long tables of numbers described in chapter II by imposing some kind of structure on them. Maybe he wasn't even aware of the method he used. Perhaps he memorized a certain chunk of numbers on each intake of breath. If Ebbinghaus did not chunk them in some way, then he holds the world's record for rote memory: memory achieved by passive, repetitive rehearsal.

But let's get back to the question about what other ways can be employed to deepen the depth of processing explanation. Consider the experience of S. F., an avid cross-country runner at Carnegie Mellon University. When initially tested with a string of numbers, he recalled seven digits—a perfectly normal performance. The psychologist working with him, K. Anders Ericsson, encouraged S. F. to practice for an hour three times a day, after which he could recite ten numbers—far less impressive than your performance when you memorized those phone numbers a moment ago.

After several hundred hours of practice, S. F. came up with a method of chunking that increased his performance to eighty digits recited at a rate of one digit per second. To get an idea of how impressive that performance is, look again at the list of numbers that you tried memorizing. That list is exactly eighty digits long. How did S. F. manage such an incredible memory performance?

Here's a hint. I mentioned that S. F. was an enthusiastic cross-country and track and field performer. This provided the key to enabling him to increase depth in processing, chunking ability, and what Ericsson referred to as "deliberate practice"—those several hundred hours of practice S. F. put in. So how did S. F. use his track and field performances to help him remember a list of eighty digits?

S. F. was able to increase his memory span for numbers by associating them whenever possible with running times from various track and field performances. For instance, S. F. encoded 3, 5, 8 as a very fast mile, three minutes and fifty-eight seconds, just short of the four-minute mile. When the sequence included four digits starting with number 3 (3, 4, 9, 3 for instance), he encoded it as 3 minutes 49.3 seconds.

S. F. was using a basic memory technique called *elaboration*. The more meaning you can give to a thing to be remembered, the more successful you will be in recalling it. Another way of doing this is to devise a sentence that can work as a code for recalling numbers. A famous example is the sentence devised by the British mathematician Sir James Jeans to represent pi, the ratio of a circle's circumference to its diameter. Here are the first fourteen decimal places of pi: 3.14159265358979.

Here is the sentence that Jeans devised to represent pi: "How I want a drink, alcoholic of course, after the heavy lectures involving quantum mechanics." Stare at that sentence for a moment and see if you can figure out how it relates to the number pi.

The number of letters in each word in Jean's sentence when read in sequence corresponds to the first fourteen decimal places of pi. The first letter is 3, the second is 1, the third is 4 letters, the fourth is 1 letter, the fifth is 5 letters, the sixth is 9 letters, etc. Running that easily learned silly sentence through your mind enables you to establish the numbers of pi, at least to the first fourteen decimal places. By using one elaboration or another, it is possible to extend the leaning of the numbers comprising pi to an astounding length. The Indian mathematician Rajan Mahadevan in 1981 correctly recited the first thirty-one thousand eight hundred and eleven digits of pi.

Perhaps at this point you are wondering, *Why should I bother? So what if I can learn to increase my digit span?*

Despite its simplicity, digit span reflects efficiency in the earlier stages of information processing within the brain. This is important because how well you learn generally depends on how efficiently you process information at the earlier stages. If you can increase your digit span, you can improve your brain's overall performance. We know this because digit span in children serves as a reliable predictor for early math and reading proficiency.

Nor should it be surprising that increases in digit span correlate with enhanced brain performance. Among the brain functions activated are attention, concentration, sequencing, number facility, and auditory and visual short-term memory.

Enormously more challenging is backward digit span. Try it for yourself. Randomly generate five or six consecutive digits on your iPhone or have someone else write down a five-digit span. As you will discover, it is much harder to do a backward digit span than it is to do a forward digit span. While many people can accurately achieve seven digits forward (the magic number 7), few can manage more than four or five digits backward. The reason for that difference? Backward digit span involves not only the registration and encoding of the numbers, but also their manipulation. You have to first envision the numbers and then "read them" from right to left rather than following the usual practice of reading the numbers from left to right.

Different brain functions are involved in such an exercise. This is a hint that we are now talking about another form of memory, the *Queen of Memory*: *working memory*. Working memory is sufficiently important that I'm going to develop my description of it in great detail later in this chapter. For now, let's stick with the forward digit span and other ways of improving it. The key to doing this is to come up with either rhyming words or images to represent numbers. First the rhyming words.

Memorize the following images that employ a rhyming word for the numbers from 0 to 10.

0. Hero
1. Sun
2. Shoe
3. Tree
4. Door
5. Hive
6. Sticks
7. Heaven
8. Skate
9. Vine
10. Hen

Thanks to the rhyming you will find the list easy to memorize. Each time you see 0 imagine a hero: someone that you greatly admire. When you see the number 7 you can imagine an angel floating on a cloud while playing a harp.

As an example of how this might work, take the number sequence 202-362-734 and memorize it using this system. You might group the first three numbers into an image of your hero, say, Tiger Woods holding a shoe in each hand: 202. In order to remember 36, picture a power saw reducing a tree to a pile of sticks. For 27, imagine an angel playing a harp shaped like a shoe. For 34, envision a giant tree falling onto a barn and smashing its door.

Now try making your own images by studying the rhyming list of one to ten. You are more likely to remember the images that you come up with yourself because in the process of composing them, you will be utilizing *association* and *elaboration,* two very basic memory techniques.

MEMORY TOOLS

Notice that the above list is based on what the numerals *sound like.* This is best for readers who learn most efficiently via the auditory mode: they learn better when listening to a lecture rather than reading it. The visual mode, on the other hand, works best by converting numbers into objects so that they look like that. Here is the list I learned from mentalist and friend Alain Nu.

1. Pen
2. Swan
3. Handcuffs
4. Sailboat
5. Hook
6. Golf club
7. Cliff

8. Hourglass
9. Monster
10. Bat and ball

This list is based on visual similarity. The number 1 looks like a pen; 2 looks like a swan; 3 joined with its mirror image looks like handcuffs; 4 looks (sort of) like a sailboat; 5 looks (sort of) like a hook; 6 looks like a childish drawing of a golf club; 7 looks like the side of a cliff; 8 looks like an hourglass; 9 looks like a lake monster pulling his head out of the water; 10 looks like a bat and ball.

If you use the second list, you are visualizing everything. Of course, the correspondence isn't perfect. The number 8 doesn't look exactly like an hourglass, but with little imaginative configuring, the match is close enough.

I suggest you try both methods to see which one works best for you. Write, don't type, each of the lists on a separate index card. By writing them down, you employ additional sensory input involving feedback from the muscles of the hand and wrist. You can't experience this when typing—an argument for writing with a pen. The employment of these additional senses can greatly expand one's memorization powers.

Some people are able to recruit additional senses such as taste or smell; other people can exchange one sensory input for another:

"When I look at a sequence of numbers, my head begins to fill with colors, shapes, and textures, that knit together spontaneously to form a visual image. To recall each digit, I simply retrace the different shapes and textures in my head and read the numbers out of them," writes Daniel Tammet, one of only an estimated fifty true memory savants living in the world today. In his autobiography *Born on a Blue Day*, Tammet writes, "As the sequence of digits grows, my numerical landscapes become more complex and layered, until they become like an entire country in my mind composed of numbers." As this quote illustrates, Tammet's memory system is partly based on synesthesia: one type of sensory stimulation evokes the sensation of another, as when the visualization of a number or a letter evokes the visualization of a color or a taste.

Let me be clear here: synesthesia is a rarity, and it's unlikely you will be able to utilize it to its fullest form a la Tammet. But in its attenuated forms we all use it in our language. We speak of "loud" ties to mean garishly colored ones; "green" with envy when we want to capture a character trait in a color; musical pitches can be imagined as dark or light and airy. The most common form of synesthesia is grapheme-color synesthesia (colored letters), with chromesthesia (music in colors) running a close second. Grapheme-color synesthesia can be learned to some extent, but not anywhere close to the degree of proficiency displayed by Tammet.

Among all of the mnemonists that I have encountered, none of them have been gifted with more than perhaps a smidgen of grapheme-color synesthesia. Instead, they make use of sight along with the other four senses (how something sounds, feels like, smells like, and tastes like) assembled according to individual talent and preference. In most cases, the clarity of one sense is heightened in comparison to all of the others.

With the exception of sight-deprived individuals, most people use the visual sense preferentially. But for some artists and others, one of the senses is accentuated by excluding as much as possible the input from the other senses. For instance, when the Canadian pianist Jen Lisiecki was asked why he played most of his concerts with his eyes closed, he responded: "When you shut down one of your senses, you heighten the others. Generally there are no smells on stage, so I heighten my hearing and completely immerse myself in the music. After so many hours spent on the piano, I pretty much know where all of the keys are."

I would suggest you focus initially on creating a visual scene in your head by closing your eyes while seeing in your mind's eye with the greatest clarity possible the object you want to memorize. Let's say you are working on creating your internal image of the sailboat for number 4. When you have achieved something that satisfies you in terms of its internal visual image of a sailboat, it is easy to imagine the sound of the wind through the sails, the streaming of the water along the hull, the tangy smell of salt water, the sense of mild unsteadiness accompanying a ride on a swiftly moving sailboat. All of this is not synesthesia, but still includes multiple senses blended together into a multi-coded sailing experience. But initially just go for that clearly visualized sailboat.

HYPERPHANTASIA AND HYPOPHANTASIA

In chapter II, we mentioned the poet Simonides and his recall of the seating arrangement in a banquet hall which had collapsed. This was followed almost two thousand years later when Giulio Camillo suggested *The Memory Theatre*, which by means of images and loci (the position of these images) vast amounts of information can be remembered. The method of loci remains as one of the most popular methods used today by mnemonists. It is the method that I personally find most helpful in committing manageable amounts of information to memory.

The first of my loci images is the front of my house, which I have studied many times during manageable amounts of information all four seasons of the year. I know the location of all of the windows. I can envision the bricks, even how many brick columns separate one window from another. Not quite the equivalent of a photograph, but close.

This need for near photographic clarity is the reason I suggest that you choose as a memory foci a personally familiar or neutral object. (In the interest of neighborhood harmony, it is probably not a good idea to stand and stare at somebody else's house for prolonged periods at unpredictable times.) So for your first loci, choose something that you can study in all of its detail whenever you want. Working initially with only one location is important because not everyone has the same capacity to form vivid images. We know this from studies on *hyperphantasia*—the ability to form intensely vivid internal images.

Working on one image will provide you insight into your visual abilities. As one researcher on *hyperphantasia* described it, the person endowed with hyperphantasia can watch a movie and later watch it in their memory and the two are indistinguishable. Despite how strange this condition might sound to you, it is not as uncommon as you might expect: 2.6 percent of the general population can relate.

The polar opposite is *hypophantasia*—the much reduced ability to form mental images. These extremes of the "mind's eye," as researchers refer to them, are not so much a disorder, as they are opposite ends on a continuum. Most people—including myself—fit somewhere near the middle of the continuum. But even if you fit somewhere close to hypophantasia, you can with practice improve your mental imagery. Finally sharpening that initial image will get you started on fashioning others and will provide insight into your own personal "mind's eye."

As we learned from studying the methods of the ancient Greeks, the first important skill to be developed in creating a superpower memory is *visualization,* not just in a vague sort of way, but seeing it with the utmost clarity. At its most developed, the visualized object and its real world counterpart should be identical. Those with hyperphantasia will, from the beginning, find this easier than their counterparts with hypophantasia. This three-word mantra can narrow the gap: "practice, practice, practice."

On a table or desk, arrange ten items in any configuration. I have a pen, a book, a magic marker, an iPad, a cellphone, my sunglasses, a book, a bottle of ink, a comb, and my watch. Select ten items of your choosing. For this I recommend you include several familiar or even prized objects. That way your initial attention will be aroused.

Study the items carefully for three minutes. Now close your eyes and practice picturing them in their specific arrangement. Can you do that? If not, start with five objects. The key is to see them as unique, easily distinguishable from one another, and yet placed in a specific arrangement. When you can do that, study them in depth, one at a time.

In this exercise I selected a pen, which I helped design with the Italian luxury company Montegrappa. It has a sterling silver barrel, an outer cap in blue celluloid; the clip is finished in 18K gold. The section above the 18K gold nib is red celluloid. After studying it carefully I place it on the table, close my eyes, and mentally picture the arrangement once again. Now the pen stands out in enhanced clarity compared to the other objects.

Next, I study the watch I am wearing today, a vintage 1948 Hamilton Barton with a rectangular two-tone dial and a separate microsecond dial at the six o'clock position. After studying the watch for several minutes, details pop out to me that I hadn't noticed before. The even numbers on the face alternate with dots representing the odd hours and the six o'clock number is replaced by the microsecond dial. As I replace the watch on the table, it too, along with the pen, stands out from the other items.

When you perform this exercise, you will notice that you use a kind of high-power mental lens to visualize the details by zooming in and out. I can do this easily with the pen and the watch, but can't do it as well with the other objects because I haven't subjected them to a similarly concentrated focused scrutiny. But I will. The rule: The clearer you can see the object in your mind's eye, the easier it is to remember. Keep in mind that clarity and detail in mental imagery is directly related to the quality of your memory.

MEMORY METHOD

Perhaps the best way of showing you how to create your own memory loci, would be for me to describe my own. As mentioned earlier, my home is the first loci followed by (2) a nearby library; (3) a coffee shop; (4) a liquor store; (5) the front of Georgetown University Medical School (which I attended); (6) the entrance to Georgetown University; (7) a well-known restaurant Café Milano in Georgetown (my favorite); (8) Key Bridge connecting Georgetown to Rosslyn, Virginia; (9) the Iwo Jima War Memorial commemorating the Marines raising the stars and stripes on Iwo Jima's Mount Suribachi; and (10) Reagan Airport. I encounter all of these sites when I walk from my home to the Iwo Jima War Memorial (approximately four miles), with the Reagan Airport nine additional miles away. I have chosen each loci based on either personal familiarity (my home, the library, the coffee shop, the liquor store, Georgetown University Medical School, the entrance to Georgetown University, and Café Milano) or by locations selected because of pictorial qualities (the Key Bridge, the Iwo Jima War Memorial, and Reagan Airport).

I suggest you choose ten locations that have meaning for you. This will enable you to start with the advantage that these locations are familiar. Because they are well known, you will start with a clear picture that can be further embellished by practice.

Here is an example of how I use these loci. Imagine I'm going to the supermarket for the following items:

Milk
Bread
Breakfast cereal
Fish
Steaks
Hotdogs and rolls
Paper towels
Watermelon
Ketchup
Orange juice

Since these items do not form any sort of hierarchy, they can be encoded via the loci in any order. Here is how I did it:

House: Imagine the house as a quart of milk turned on its side with the milk pouring out of the chimney.

<u>Library</u>: When I look through the floor-length window facing me, I see loaves of bread instead of books on the shelves.

<u>Coffee shop</u>: A giant coffee cup on a table outside contains a watermelon   inside it.

<u>Liquor store</u>: Hotdogs and buns inside bottles lining the shelves.

<u>Georgetown University Medical School</u>: A box of cereal.

<u>Georgetown University Main Entrance</u>: The entrance is flooded in ketchup with the students wading in knee-high rivulets of ketchup.

<u>Café Milano</u>: The restaurant is under water and schools of fish flirt while suggestively flitting from bars to tables and back again.

<u>Key Bridge</u>: The surface consists of a prime steak being tenderized by cars traveling back and forth from Georgetown to Rosslyn and vice versa.

<u>Iwo Jima Memorial</u>: The Marines (with apologies, no irreverence intended) are hoisting a giant paper towel.

<u>Ronald Reagan Airport</u>: The enlarged statue of our fortieth president Ronald Reagan at the airport bearing his name and featuring the president holding a glass of orange juice for the thirsty traveler.

In order to get started, come up with your own memory loci consisting of location you frequently encounter every day when you walk in your neighborhood, objects associated with your typical day starting with waking up (bed) and ending with the last thing you do at night (turning off your beside lamp).

You don't have to limit yourself to one series of loci. Along with my home loci, which I used in the example above, I have another one made up of objects and places encountered when I get up from my desk at the office and walk to an Italian restaurant two blocks down the street. Anything will do, as long as you select as loci any group of objects that appeal to you. The important thing is that the loci become second nature to you; you can instantly remember them and see them with the utmost clarity.

Remember, the more bizarre or irreverent the images, the easier it is to remember. But the primary determinant of your success, as mentioned throughout this book, is the clarity of the images. Continue to work on seeing the loci that you are going to use merge with the information that you are trying to remember as clearly as you possibly can.

THE ZEIGARNIK EFFECT

"But when I use the method of loci on one occasion, doesn't that interfere with my use of it again?" you may ask. In other words, don't the images from one trial (like the grocery list we just completed) mix with those from another memory list and thereby make memorization harder? That's a good question and I wondered about it myself until I discovered the Zeigarnik effect.

In the 1920s the Soviet psychologist Bluma Wulfovna Zeigarnik set out to explain why a waitress with an extraordinary memory for food orders never had to write anything down and promptly forgot the orders after the customer had paid and left the restaurant. To find out what was going on, Zeigarnik asked volunteers to carry out various tasks, such as forming a clay figure, working on a math problem, or assembling a cardboard box. Before they finished these tasks, Zeigarnik interrupted them and established what they remembered about what they had been doing.

Zeigarnik discovered an oddity of the human brain: when a task is left uncompleted, we tend to compulsively keep thinking about it. This prevents us from forgetting it (90 percent better recall for interrupted versus completed tasks). In the cafe example, when the customer had eaten and paid the bill, a certain mental tension was released in the waiter (no more possibility of screwing up an order or being drawn into a disagreement over the final tally).

Memory, along with the memory loci we just discussed, conform to the Zeigarnik effect. When you complete your supermarket purchases—stored as images on your memory loci—the linkages will greatly weaken and soon be forgotten altogether. Your memory journey (from, in my case, my home to the Reagan Airport) will continue to be available. All you have to do is frequently review your loci circuit.

RESTAURANTS AS MEMORY LABORATORIES

Memory loci don't have to be extensive like mine (ten loci located along the thirteen-mile route from my house to Reagan Airport) but can be confined to a single room. In some ways small "memory theatres" are easier, since they provide tightly knit spaces. In other ways they are harder to work with because the loci may overlap, thus obscuring the images you must remember with absolute clarity in order to effectively use them. Though I wouldn't recommend it, sometimes only one loci can be employed.

For instance, a waiter with a reputation for never messing up an order explained to me how he did it: "First thing I do when I get to work is to memorize all of the day's specials. The remainder of the menu doesn't change much from day to day and I've already memorized it. When a customer orders a meal, I mentally superimpose their face on the menu alongside their order. Finally, as a backup I envision the words comprising the order as coming out of their mouths like a thought bubble that you see in a cartoon. When I'm in the kitchen I mentally review the faces-menu-orders images. If I forget one, I have only to look out at the table and the face will cue the order." In this method, a single loci (the menu) provides the waiter the background upon which both faces and food orders appear.

Another restaurant specialist, Jacques Scarella, served for many years as a maître d at a restaurant in Washington, DC. At the time I interviewed him, he had spent over forty years in the restaurant business. A dapper trim man with silver-grey hair, he was able to remember specific dinners favored by his customers, some of whom he hadn't seen in the restaurant for several years. In the morning after a busy evening in the restaurant, Jacques could picture the forty-six tables in terms of the customers, as well as their exact table and place settings.

In response to my inquiry as to how he was able to accomplish this memory feat, Scarella turned out to be a rare example of someone with a super-power memory who didn't rely on a memory system. "It's not just the orders that I remember. I can recall a customer's attitude during the time they were here last. When they order, I can help direct their attention to those things which in the past have given them pleasure."

When Scarella is asked about his deserved reputation as the possessor of a super-power memory, he attributes it to a "natural gift" developed in tandem with an insatiable desire to succeed. To him it's all part of his European-learned-credo as a restaurateur: the customer's preference is important and therefore should be remembered.

I mention Scarella here because I want to re-emphasize the importance of *attentive observation*. If the focus and desire are raised to a sufficient level, the use of a memory system may not be necessary. But one caveat is important if you consider doing that: You are unlikely to achieve anywhere near your memory potential without a system. For instance, Scarella's super-power memory is effective only within the restaurant domain. I learned that by asking him the street intersection at the corner closest to his house. He couldn't tell me. "Why should I bother to remember it? I know how to get home."

THE QUEEN OF MEMORY

Previously, I dubbed working memory as the "queen of memory." I did so because working memory is the most important enhancer of intelligence. When you strengthen working memory, you coincidentally increase your intelligence.

Working memory is essentially the ability to keep in the attentional foreground a piece of information while you turn your attention to something else. If while I'm writing this section of my book, my wife interrupts me with a question, I may discover after a brief discussion with her that I have forgotten what I intended to write next before the interruption—one of the reasons writers don't like to be interrupted while writing, unless for emergencies. This "interference effect" results from a failure in working memory. In this case, I failed to keep in mind the point that I was intending to make in the paragraph I was writing at the moment I turned my full attention to my wife's request. As a result, when I then returned back to writing I was unable to retrieve what I intended to write before the interruption.

As the form of memory most highly evolved in humans, working memory involves the active storage and manipulation of information. Mental arithmetic is a good example. When mentally multiplying two-digit numbers, it is necessary to add the sums from each of the digits and then add them together. Some people have great difficulty doing mental multiplication and require either pen and paper or a calculating device. The demand on working memory is even greater when multiplying three-digit numbers and remains beyond the capacity of most people, even though the mathematical process is exactly the same as with two-digit numbers. The increased difficulty with three-digit multiplication stems from the increase in information that must be actively stored and manipulated in working memory in order to come up with the final product.

The more you can remember of the events of your life, the greater your chances of making unexpected and liberating personal insights. Speaking metaphorically about the power of memory, Japanese novelist Haruki Murakami has written "People's memories may be the fuel they burn to stay alive." Murakami is referring principally to working memory. Enhancing working memory in childhood is important since deficits in working memory lead to poor school performance, especially in reading and mathematics. That's because, absent an efficiently functioning working memory, it's hard to follow detailed instructions, mentally manipulate numbers, or process lengthy passages of text. In computer terms, it's as if the passages encountered earlier in the text cannot be held "online" and integrated with what comes later.

Thus working memory serves as bottleneck for taking in information and storing it. When one cannot integrate what has been heard and read only moments earlier with what one is currently encountering, learning and the subsequent establishment of memory are severely compromised. That's why the quality of a person's memory has a lot to with their success in life. For one thing, intelligence—as measured by standard IQ tests—is associated with a finally honed working memory; in general, the smarter the person the more efficient their working memory. Isn't this reason enough to make efforts aimed at increasing your working memory?

Working memory is often metaphorically compared to juggling. A good juggler keeps within his attention span and memory a varying number of balls in the air. Working memory is like that. It involves a relatively small number of items (averaging three or four for a visual working memory) that are simultaneously kept track of. Just as some people can juggle more balls than other people, the numbers of items for working memory varies from one person to the next.

For reasons that aren't completely understood, but illustrated by the Zeigarnik effect mentioned earlier, the brain tends to remember uncompleted or interrupted tasks better than completed ones. So if you want to remember something, it's best to take a short break before you finish reviewing it. By temporarily turning to some relaxing diversion (a short phone call, conversation etc.), you will wind up with a stronger memory for material than you would if you plodded on without a pause. Psychologists speculate that during the process of establishing a memory the brain maintains a creative tension, which is only relieved when the learning process ends and the memory is established. Because this tension is continued during your break, your memory for material reviewed both before and after the break will be more readily accessible, or easily learned and more successfully remembered.

Another highly effective technique for improving your memory is to keep retesting yourself on the material you want to remember. Even after you have learned something, your long-term memory for it will be strengthened if you repeatedly challenge yourself to recall it again and again.

In one experiment establishing this rule, English-speaking students learned pairs of Swahili and English words. For example, if they were given the Swahili word "Mashua" they had to provide the correct response "Boat." Since the students had no previous exposure to Swahili, they couldn't rely on background knowledge to help them establish a memory for the Swahili-English combinations. One group retested themselves repeatedly in all of the word combinations including those they had successfully identified on previous testing. Another group stopped testing their memory for word combinations once they had identified them correctly on the test.

The memory difference between the two groups on their final examination was dramatic. Those who stopped testing themselves for word combinations once they had correctly identified them on an earlier test, could only remember about 35 percent of the word pairs compared to an 80 percent remembrance rate for those who kept testing themselves for all the word pairs throughout the experiment. Bottom line: The more times an item is retrieved from memory, the more likely it is that the term can later be recalled. Frequent retrieval, it turns out, is even more effective in establishing a memory than relying on additional studying.

Now let's reconsider what happens when someone tells you a series of actions to be taken at work. You won't be able to do them unless you keep in mind the correct sequence from start to finish. What you are doing essentially is encoding one item while retaining access to items encoded moments earlier. This is referred to as *working memory in action.* Experts consider it the basis for reasoning. In general, those people who can hold the greatest numbers of items in mind are best at considering multiple aspects of a problem simultaneously.

Word processing on your PC provides a good analogy for what happens during failures of working memory. As you switch from document one to document two on your word processor, the unattended document (one) is still accessible to you. All you have to do is toggle from document one to document two in order to keep both of them "in mind." But if you close the first document as you move to the second document, that first document will no longer be available for redisplay. A failure in working memory is like that: you close the first document in your mind and you switch to another instead of holding that original document online.

THE N-BACK GAME

Here is a working memory exercise you can carry out at leisure. Memorize these numbers: 270183. Reading the sequence silently to yourself once or twice should do it. Now, mentally rearrange that sequence of numbers in your head from lowest to highest without writing anything down. When you have done that, rearrange the sequence from highest to lowest. The only way of carrying out these two exercises is to mentally envision the original sequence and then mentally manipulate the numbers.

Here is a demonstration of working memory using a deck of cards.

1. Shuffle the deck and place it facedown on the table. This is the draw pile.
2. Choose two cards—say, a deuce and a queen—to be trigger cards. Now turn the cards over one at a time from the draw pile and place the card on a discard pile. Only one card should be face-up and visible at a time.

3. Whenever you draw a trigger card (a deuce or a queen),
   name the card turned over two turns earlier. Since the face
   of this card is now hidden in the discard file, you must use
   your working memory. If your selection—an act of working
   memory—is correct, continue drawing cards until you
   encounter another trigger card and once again try to remember
   the card that was two cards back (the two-back challenge).

As you try this working memory exercise, you'll notice a different
kind of mental exertion is required than with the earlier testing of
immediate memory. You will have to hold two cards in memory at all
times. This is a much more challenging, and *distinct* type of mental
pressure than you experienced when you simply remembered all the
numbers in a string of numbers.

Your brain must keep in mind several operations in order to succeed at the two-back challenge. When you encounter the trigger
card, you have to remember the card that you experienced two cards
earlier. In the three-back challenge, the goal is to remember back
three cards. After you become skilled at two-back, try three-back or
perhaps even four-back. You can go back as far as you dare. That's
why this class of test is referred to as N-back. N can be any number.

There is also an auditory method that you can use to play N-back.
You will need to use either a recording device or your cell-phone.
Record the following sequence of letters aloud at a rate of about one
letter per second:

PRFBA FALLP BRADB LYIYI ULYYU URUNY
FIPAL LFIUR PLAYF LINUY.

After you have recorded them, take a short break and occupy
yourself with something else. Return fifteen minutes later and play
back the recording while listening for a randomly selected target letter (B, for instance). When you hear that target letter, stop the recording and write down the letter that you heard two letters previously.

When you are finished, pick a new target letter (A). When that target letter comes up, repeat the process of writing down the letter that came up two letters before. Check yourself against the written sequence. You can then move to three-back.

Another challenging test of working memory can be done with coins. Gather a handful of nickels, dimes, pennies, and quarters and place them in front of you on a table. Don't count the number of coins ahead of time, but the desired number should be somewhere between five and ten of each, laid out in no particular order. Now pick them up one at a time and total them. How did you do that? Unless you have a highly developed working memory, you picked up the coins randomly and kept a total in your mind as you processed them. Most people have little problem doing that exercise.

Now pick them up one denomination at a time and total it before moving on to another denomination. Pick up all the pennies, total them in your mind, and then the nickels, total them, etc. It's easiest to count the coins one denomination at a time because that process makes less of a demand on working memory. After counting the nickels for instance, that total can be stored in working memory and your attention shifted to the next denomination. At the end you have only to total all of the denominations. This is only a slightly more difficult working memory exercise than the previous one.

In order to really challenge your working memory, start with two coins such as pennies and nickels and count them at random (i.e., don't alternate them). This will require you to keep a running total of each denomination in working memory while you are counting. Do it rapidly as you count each coin and discard it by placing it to the side. When you are finished, write down your separate totals and then check for accuracy by counting each of the denominations among the discarded coins. The mental calculation and the sum of the discarded coins should be the same. Not much practice should be required for you to manage the two denominations.

Next, count three denominations simultaneously such as pennies and nickels and dimes. Then simultaneously count four denominations. If you can manage four, you are achieving the maximum number of items that can be worked with using working memory. It's a fundamental characteristic of working memory, that "four items seems to be the limit" in this exercise, according to Paul Verhaeghen, the psychologist who carried out the experiments demonstrating the four-category limit. He discovered this and published it in a fascinating paper "People Can Boost Their Working Memory Through Practice" (see Bibliography), published in the American Psychological Association's *Journal of Experimental Psychology: Learning, Memory and Cognition (Volume 30, No.6)*.

Work at any of the above suggested exercises frequently. They don't require any expensive or fancy equipment: playing cards, a voice recorder, and a handful of coins ranging from pennies to quarters. By practicing these exercises, you will become familiar with the principles involved in testing and strengthening working memory. At that point you will be in a position to fashion your own working memory challenges.

A specific area of the brain is principally involved in working memory: the prefrontal cortex, the most anterior (farthest to the front) portion of the frontal lobes on each side of the brain (the parietal lobes also play a part). In humans, the frontal lobes are among the last structures to mature; they encompass almost a third of the entire cortex. Localized portions of the prefrontal cortex (the dorsolateral prefrontal cortex), is especially involved in working memory.

As an example of the prefrontal cortex in action, consider what happens when I'm looking for a favorite pen that I can't find. As soon as I realize that it has gone missing, I begin looking for it in a more or less systematic manner, starting with the location of where I remember last seeing it (on my desk). When I don't find it there, I go through several briefcases, search the inside pockets of several sport coats, etc. If I don't find it in those places, I widen my search, taking care not to revisit places I have already searched. Throughout my quest I remain focused on the goal of finding the pen and ignore any distractions. It's the dorsolateral prefrontal cortex that is assiduously working during my search for the pen. As an aid, it also fashions an internal representation (mental snapshot of the pen), which helps me to remain focused during my search for it. Here we have a quintessential example of working memory in action.

Children and adolescents possess underdeveloped prefrontal cortices—they easily lose things and experience difficulty finding them. Their brains aren't fully matured until early adulthood—as a result, their working memory isn't very efficient: they experience problems organizing themselves, keeping their attention focused, and managing more than one or two things at a time.

In old age, especially among people with degenerative brain diseases of the frontal lobes, working memory may suffer a decline. But such a decline is not inevitable. Indeed, it is one of the basic tenets of this book that this degeneration of the frontal lobes that accompanies aging can be slowed, even if not entirely stopped, by working memory exercises.

TAMPING IRON MEETS BRAIN

If you had to select one area of the brain most important for those mental qualities that distinguish the working memory of humans from other living creatures, it would be the frontal lobes.

All of the memory centers connect with the frontal lobes, which are concerned with judgement and imagining of the future: if the frontal lobes are damaged secondary to trauma or degeneration, the brain areas responsible for memory become untrustworthy. Someone with frontaltemporal dementia (the disease that killed Robin Williams), for instance, cannot reliably access their memories. Thus they may lose some or all of their memories secondary to destructive effects on the frontal lobes. They don't always suffer from memory loss exactly like a person with frontal damage secondary to Alzheimer's disease, but sometimes they do. In those cases the loss of memory is related to a direct attack on the memory centers (primarily hippocampus and amygdala), which form connections with the frontal lobes.

Here is a list of the principal activities mediated by the frontal lobes:

Drive/ motivation:	Frontal lobe damage results in the loss of ambition and self-motivated behavior. External sources of stimulation and inspiration become more important as motivators than internal self-direction.
Sequencing:	In frontal-lobe damage the person lacks the ability to keep information in proper sequence. He also has difficulty in separating the most essential information from less important background material.
Executive control:	Planning and anticipating the consequences of behavior are disrupted with frontal lobe damage. In some patients this loss of self-monitoring extends to social behavior. They make crude and insulting comments, tell obscene stories, or openly express sexual interest only moments after meeting someone. They exercise poor judgement and generally lack the ability to see things from other people's point of view.
Future Memory:	Almost all complex human activities require some advanced planning. This involves imaginatively comparing how things are now and how one wishes them to be. The resulting internal model of the future serves as a guide for altering and updating one's behavior in the direction of achieving the goal. This way the individual is acting on what we refer to as a "memory of the future."

Self-analysis: As a result of any disruption in the sense of self-continuity, recognition is lost of a stable self that extends from past through the present into the future. The person loses self awareness as a changing, evolving person with personal responsibility for his own actions. He can no longer project himself into the future or integrate experiences from the past.

The classic example of a loss of frontal lobe influence on working memory and executive function dates from a railway accident in 1848 outside Cavendish, Vermont. A railway crew was about to set off an explosion after drilling a hole in a rock and filling it with gun powder. When this was completed the foreman, Phineas Gage, picked up a tamping iron and pushed it down into the hole. The tamping iron prematurely set off a spark when it scraped against the sides of the drilled hole. As a result of the ensuing explosion, the tamping iron shot with the force of a cannon ball and struck Gage beneath the left eye and tore through the frontal lobe and frontal bone of his skull, exiting just above the hairline. Miraculously, Gage *physically* survived this horrifying trauma (his crew thought he was dead when they observed him catapulted into the air and landing on the ground).

I qualified Gage's survival as *physical* because he was a totally changed man. Here is the description by his physician John Harlow, who attended Phineas Gage for several years:

"He is impatient of restraint or advice when it conflicts with his desires, at times pertinaciously obstinate, yet capricious and vacillating, devising many plans of future operation which are no sooner arranged than they are abandoned in turn for others, apparently more feasible. Previous to his injury, he possessed a well-balanced mind and was looked upon by those who knew him as a shrewd, smart business man, very energetic and persistent in executing all of his plans of operation. In this regard his mind

was radically changed so decidedly that his friends and acquaintances said that he was "no longer 'Gage'."

As a subtler example of what can go wrong with the frontal lobes and what effect this has on working memory, consider the case of David, who suffered frontal lobe damage as a result of a rupture of a blood vessel near the frontal lobes. After surgery and rehabilitation, he returned to his job in personnel management. After several weeks of working, he was sent to my office with a letter from his supervisor detailing unfavorable changes in his personality and behavior.

Although he was formerly "a self-starter," David now required almost constant supervision and had to be prodded into performing even routine office tasks. When at home, he spent almost all of his time watching television, rarely read anything, and had little to say, preferring to stay inside rather than, as was his former style, visiting with friends. On testing he could quickly learn and recite a short shopping list but forgot most of the items after a short delay, or if distracted. "He is highly dependent on the presence of context in his ability to encode and store verbal information," according to the psychological report. The report continued, "Unless he is told specifically what to enter into the systems, he seems to have a great deal of trouble focusing his attention on the job at hand. He is continuously printing unnecessary copies of every screen he has on his terminal. He buries himself in paper, has trouble weeding out unnecessary data. He has difficulty keeping sequential steps in order. Without external intervention he leaps from task to task lacking a sense or understanding or purpose. He gets side tracked by all the papers he is looking through."

Despite his difficulties, David performed normally on psychological testing. It was only when he was required to keep in mind his goals at the moment (organizing work information on his computer) that he couldn't balance the information or keep more than one thought in mind at a time. Working memory was seriously damaged. He also lost his drive and motivation secondary to his inability to balance circumstances and future goals simultaneously. If we were asked to come up with two words describing David's impairment it would be loss of *executive control.*

David represents an attenuated form of the frontal disorders of working memory and executive control that affected Phineas Gage.

THE LADY GAGA TICKETS

Executive control involves the planning and the anticipation of the consequences of behavior. Neither Phineas Gage nor David could maintain their performance and therefore adhered to unsuccessful approaches. Most important from our point of view is the lack of an intact working memory. David could not maintain more than one object in mind at a time. Contrast David with that of a successfully functioning executive.

The executive of a corporation is responsible for productivity, stock holder satisfaction, and worker compensation along with making various micro-decisions every day. All of these elements have to be carefully kept in mind and balanced. In order to remain balanced they have to be kept in working memory. When you strengthen your working memory, you enhance your life skills by keeping in mind past performances, your present situation, and your future desires. Sadly, my patient David, like Phineas Gage a century and a half before him, had lost this ability, all of these secondary to the damage to his frontal lobes.

Let's take a closer look at executive function and how it helps organize our memory and our life. Think back to a typical day in college. Just imagine you were scheduled to take a French class at 10:00 a.m., a chemistry class at 3:00 p.m., and a public speaking class at 4:00 p.m. You also needed to do laundry, pick up tickets for a Lady Gaga show that evening, withdraw some cash from your bank account, and, finally, be primped and ready for a dinner date at 6:00 p.m.

In order to accomplish these goals, you have to keep them all in mind and organize them. Some can be done anytime and do not impose a specific time constraint, (doing the laundry, picking up tickets for the Lady Gaga show, withdrawing spare cash from the bank). But everything has to be done by no later than 5:30 p.m. (after all, you need *some* time to prepare for the date).

To accomplish all this you have to mentally envision the proximity of your classes to the laundry, the concert ticket pickup spot, and the bank. And since each of your classes are held in different buildings across campus, efficient planning is required, lest you run late and turn into a no-show for the 6:00 p.m. date.

In carrying out this exercise in temporal and spatial planning, you depended on the central executive and working memory to organize and achieve all of these goals in the most efficient manner. You have to remain consciously aware and review them in your mind via inner-talk ("Isn't the laundry a half a block from the French class? So I'll do these together") and mental pictures ("Yes, it is. I can see them both in my mind's eye"). The central executive and working memory are cooperating in tandem. Indeed, without the cooperation of working memory and central executive, you wouldn't be able to carry out the required mental juggling.

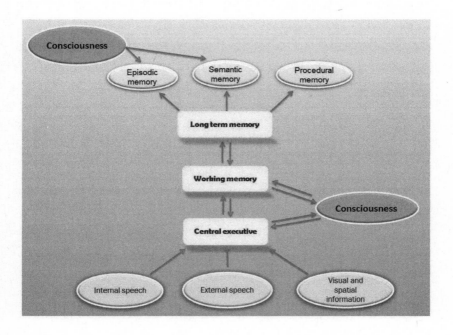

Notice in the above graphic the two-way communication between central executive and working memory. This makes anatomic sense, since each resides in the pre-frontal cortex. That's the reason why damage to this area located toward the very front of the brain results in defects in both working memory and executive function. Remember what happened to Dave when he sustained pre-frontal brain injury.

"NEGATIVE TWENTY QUESTIONS"

In the final pages of this discussion on working memory, let's explore additional exercises that require no special materials and can be done at any time and any place.

Go over in your mind the US presidents backward from Joseph Biden to Franklin Delano Roosevelt. Write them out, or speak them in order using your voice recorder or cell phone. Now name them or review them in your mind from Roosevelt to Biden. Next, name only the Democratic presidents, skipping the Republican presidents; then name from memory the Republican presidents while skipping the Democratic office holders. Now name them in alphabetical order regardless of party. Notice that in the later two operations (ordering by party or alphabetical order) you have to *work harder*—you are engaged in a very challenging working memory exercise.

Next, list in your mind the members of your favorite sports team. Name them along with the positions that they play. Now without writing anything down, list the players in alphabetical order. Finally, name them in reverse alphabetical order.

If you are not a sports fan, write down (writing is ok here because you have to make up the list before carrying out any manipulations with it) your ten favorite authors. Memorize that list (immediate memory), which shouldn't be difficult since they are *your* favorites. Now mentally list them in alphabetical order. If you forget one, you are allowed to take a one-time brief glance at the list. Finally, list them in chronological order according to the years during which they were written. In the case of overlap (Hemingway and Faulkner, for instance) list their names in alphabetical order (Faulkner, Hemingway).

While doing any of these exercises, you have to maintain information and move it around in your mind, according to superimposed criteria—the essence of working memory.

Reading novels, incidentally, provides an especially helpful exercise in working memory. Why is fiction preferred over non-fiction? Because non-fiction works are often organized in ways that allow the reader to skip around a bit according to personal interests and previous familiarity with the subject (you are now reading such a book). Fiction, on the other hand, requires the reader to proceed from beginning to end while retaining in working memory the various characters and plot developments.

The fiction reader must remember when a character was first encountered along with all of the backstory from the character's past. Each time that character reappears, the reader, by means of working memory, has to recall the character's past actions along with whatever insight into the character's motivations the author may provide.

Incidentally, I have noticed over my years as a neurologist and neuropsychiatrist that people with early dementia, as one of the first signs of the encroaching illness, often stop reading fiction. They can no longer keep the characters or plot development "in mind" (in their working memory). A second early sign of incipient dementia (while we are on the subject) relates to cooking. Unable to retain and employ working memory, the sufferer can no longer follow a recipe. Especially hard are measuring the ingredients and timing their entry into the meal being prepared. Bottom line: keep reading and cooking as spurs to maintaining your working memory.

Games offer another avenue for strengthening working memory. Bridge and chess are stand-out examples of keeping past, present, and future memories (based on evaluations of past games and the future consequences of the decisions made during those past games). My favorite working memory game is Twenty Questions.

In the traditional game of Twenty Questions, one person, known as the questioner, leaves the room and the remaining players select a person, place, or thing. When the questioner returns to the room, he or she tries to guess what the group has decided by asking a total of no more than twenty questions.

According to the rules, the questioner can operate on the assumption that his questions will be answered truthfully and that the unknown person, place, or thing that he is trying to guess does not change. The skill demanded of all the participants is to keep in their working memory all of the previous questions and answers. For instance, if the object is a "camel" and the questioner asks, "Is it an animal that lives in the ocean?" the "No" response must be kept in working memory by all of the other questioners when presenting their questions.

My version of Twenty Questions can be played by two people. You will therefore need a partner in order to play the game, and the working memory benefits accrue to both players. The rules are simple:

1. One player thinks of a person, place, or thing.
2. The second player asks as many questions as necessary— up to a limit of twenty—in order to guess that choice.
3. If the questioner by means of insightful questions guesses what the other player has in mind, he is the winner. If he fails to come up with the identity of the object after twenty questions, then the first player is declared the winner.

Success in Twenty Questions depends on the questioner's ability to keep clearly in mind all of the answers and mentally eliminating possible choices on the basis of these answers. The skill of the other player (in a two-person version) or group depends on the ability to come up with difficult-to-guess persons, places, or things. Special versions of the game can be played involving special interests—say, history or medicine—and limiting the answer to a person, place, or things taken from historical or medical sources.

A devilishly clever form of Twenty Questions was first described by quantum physicist John Wheeler. The Wheeler version, called "Negative Twenty Questions," is much more demanding in terms of working memory. Let me explain the game in terms of multiple players rather than the two-player version—which, if you decide to play that version, I'd suggest you play only with someone who is *very smart*.

Unbeknownst to the questioner when he or she leaves the room, the other players purposely omit choosing a single person, place, or thing. So when the questioner returns and begins questioning, there is no definite person, place, or thing to be guessed. Instead the person questioned thinks of a person, place or thing consistent with the question asked. "Does it live in the ocean?" "No." Each of the other players will do the same thing: after hearing the question, they will create a person, place, or thing of their own as long as it is consistent with previous questions and their answers. In this version of Twenty Questions, there is no agreed upon person, place, or thing to be guessed.

As Wheeler explained the process to science writer John Horgan, "The word wasn't in the room when I came in even though I thought it was. Not until you start asking a question do you get something. The situation cannot declare itself until you have asked your question. The asking of one question prevents and excludes the asking of another."

Wheeler applied this game of Negative Twenty Questions to quantum physics. "Every particle derives its function, its meaning, its very existence entirely—even in some context indirectly—from the elicited answers to yes-or-no questions.

In this version of the game—a kind of test for communal working memory—all of the participants must remember each of the questions and the answers. During the course of the game these questions and answers must be maintained and shifted around in working memory. If even one question and response goes missing from working memory, the game falls apart. Didn't I tell you it was devilishly clever?

"SMOOTH OPERATOR" TO "CLUMSY KLUTZ"

Prior to writing this sentence I brewed a pot of coffee. I didn't have to consult any sources, bring anything in particular to consciousness, or concentrate very forcefully on what I was doing. After preparing hundreds, probably thousands, of pots of coffee, the whole process has become automated into *procedural memory*.

If you drove your car today, your performance was pretty much automatic (unless you are in the process of learning to drive). If you are a veteran driver, you can do all kinds of things while driving that a novice driver can't: listen to the radio, converse with a passenger or, if you dare, call to make a date on your cell phone. If I ask you how you can do these things and still drive, you can't really put into words the steps leading up to your improved driving performance because it is based on procedural memory.

Procedural memory (also known as skilled memory) doesn't involve speech or words at all. Other examples include physical skills like skiing or riding a bike. After sufficient repetition of these activities the brain forms neuronal networks that spring into action whenever we are skiing or biking. After sufficient experience, these activities can be performed outside of conscious awareness. But even after we become skilled at these things, we can't explain to anybody else how to do them because procedures rather than words are involved. This is why procedural memory is sometimes mistakenly called muscle memory, Actually, the memory resides in the brain and not in the muscles. But this doesn't imply that our muscle movements play no part in how we establish memories.

Psychologists have shown, for instance, that it is easier to establish a memory if we accompany an appropriate bodily movement of what we want to remember. For example, if you were requested to memorize and later recite a series of simple requests (pick up a pencil, open a jar, light a candle, etc.) your ability to remember the requests would be greatly improved, if at the time of memorization you actually carry out these actions using a pencil, a jar, and a candle. These movements establish the required motor-muscle program underlining the instruction.

Getting back to the driving situation, if I ask you to teach me how to drive, you will have trouble doing that by verbal instruction alone because procedural memory doesn't require consciousness and is a non-declarative memory. You cannot teach someone just by words; you have to convey to them the motor program. No one has ever learned by verbal instruction alone how to drive a car, ski, play a guitar, or any skilled action.

Sometimes described as "knowing how" as opposed to "knowing that," procedural memory is an acquired sense of how to carry out a complex pattern of behavior. I can instruct you how to make a cup of coffee starting with the selection and grinding of the coffee beans, but it is much easier for me to just make the coffee myself and serve it to you. The development of any skill or habit such as coffee making involves motor action involving automatic behavior. With sufficient practice, all of the necessary actions can be carried out without consciously thinking about them. As a result, once they've become automated, they are very difficult to teach to someone else.

While procedural memory is a form of working memory, it is processed differently. The components of the action must be practiced frequently. But merely repeating a certain behavior does not guarantee the development of a skill: that perfect cup of coffee.

As a first step of practice, the desired skill is broken down in its component parts accompanied by an understanding of how these parts come together. Unnecessary or ineffective actions are omitted, as the coffee-making procedural memory is established. As a key component, important steps are accentuated and the process is sped up and automated. These steps are 1) making an attempt (preparing the coffee), 2) analyzing the results (tasting), and 3) changing the procedure, if necessary, in order to obtain the desired result. With each rehearsal of the coffee-making protocol, a neural network is established outside of conscious awareness. Feedback (access to knowledge of the result), is the key to perfecting procedural memory: test the coffee yourself (internal feedback) and take into account the opinions of others (external feedback).

Think of procedural memory not only as a form of working memory involving the muscular system. It also involves different parts of the brain, primarily a group of structures below the cortex called the *basal ganglia*, which includes such components as the stratum. As skills further develop, changes in the cortex also occur—the motor cortex, for instance, is linked to the learning of the new motor skill. Deep in my own brain is a cortical, as well as subcortical signature for my coffee-making ability. Elsewhere in my brain are the signatures for riding a bicycle, swimming, or dancing, and these involve different motor programs.

The automated performances of procedural memory provide both promise and peril. Once something has been practiced and learned to the extent that conscious awareness isn't necessary, reintroducing conscious awareness can prove an extreme liability. Once you have learned that new dance step, you better resist the temptation to further improve it by becoming aware of your feet movements, lest you segue from a "smooth operator" to a clumsy klutz.

THE TWISTIES

Procedural memory can also go afoul for professional athletes who have a name for it: *choking under pressure*. In response to mental pressure the athlete shifts, sometimes for only a few seconds or so, from performing to analyzing the performance. A combination of performance anxiety, increased self-consciousness, and stress about performing at the highest level possible, leads to an increase in the attention paid to the process involved in the execution of the skill. By shifting conscious attention to the step by step components, the athlete disrupts the automatic (proceduralized) performance. What was formerly a smooth unconscious retrieval of a procedural memory becomes a conscious and troubled one.

As an example, Simone Biles, considered by many the greatest gymnast in the history of the sport, was the all-out favorite to win a gold medal at the 2021 Tokyo Olympics. In the final event Biles was prepared to perform a two-and-a-half twisting vault, but she froze after just a one-and-a-half twist. This freezing is known as the "twisties" among gymnasts. In comparisons to other sports, choking under pressure in gymnastics can result in serious injury, even death. "You have to be there 100 percent or 120 percent because if you are not, the slightest bit, you can get hurt," says Biles.

Imagine yourself twisting through the air when you have the momentary sensation that you have lost control of your body. In response, you might twist at a point you hadn't planned to, or you might do as Biles did and simplify your performance by doing a one-and-a-half rather than a two-and-a-half twist.

Recently, neuroscientists have learned that choking under pressure occurs in other creatures than ourselves. A study from Carnegie Mellon and the University of Pittsburgh shows that monkeys can learn to anticipate rewards based on visual cues shown on a computer screen. Different colored objects correspond to different levels of reward (smaller to larger sips of water). When the anticipated reward was suddenly doubled or tripled, the performers improved their performance—but only up to a point. When a really big reward was anticipated, the monkey failed.

To find out the basis of choking, the monkey's movements were found to separate into two phases: a fast initial "ballistic reach motion" to direct the cursor to the target, followed by a slower more precise movement aimed at landing precisely on the target. In the choke situation, the normally fast ballistic movement stopped short of the goal. This resulted in a prolonged "homing step" that dragged on until the time ran out. In a phrase, the monkey's choking was based on overcaution. They paid too close attention to their movements (explicit monitoring), and this over-attention led to slower not faster movements. "The monkeys are psyching themselves out and under-shooting," according to Aaron Batista, a bioengineer involved in the study. To put the monkey's performance in human terms, choking resulted from an *increase* in caution.

While the Carnegie Mellon research doesn't pinpoint the exact brain circuitry involved in choking, future studies using brain-im-planted electrodes should nail down the circuitry responsible for overcaution based choking. But for our purposes, the Carnegie study points to the importance of not controlling our procedural memory from the top. In order to achieve the best performance we are capa-ble of, we have to stop micromanaging our body.

"Let the brain be the brain" is the motto that best prevents chok-ing under pressure. Once a motor act is refined to the point you can do it without effort, do just that. Perform now and analyze later, perhaps with the assistance of a coach. As one athlete suggested to me as an antidote to choking, "Plan tight, play loose."

THE FORGOTTEN BABY SYNDROME

There are other potential perils when procedural memory and con-scious deliberate memory come into conflict. On occasion procedural memory, like a stealthy burglar, can sneak in and displace deliberate conscious memory.

For instance, a subtle but remarkably effective tennis ploy involves making the opposite player conscious of a usually unconscious auto-matic maneuver (part of their procedural memory). "Hey, that was a great serve. Let me see how you grip the racket." Suddenly the opposing player, if not aware of the ploy, is tricked into consciously describing or demonstrating something that had been automated and working well. The request for a demonstration of the hand on the racket converted a non-declarative procedural memory to a less efficient declarative one. Typically—and for reasons that remain unexplained, except to those acquainted with this nefarious little trick—the tennis match turns around after the demonstration.

Sometimes a conflict between procedural memory and consciously mediated memory can lead to disastrous consequences. For example, a London bus driver with a twenty-year history of driving single-decker buses was switched to double-decker buses. Since this change carried potentially serious consequences, the driver underwent full retraining, starting with his attendance at classes designed for new hires assigned to double-decker buses. The result? Everything went great for two years until one harried afternoon.

The driver, running late, allowed his procedural memory to kick in and took a shorter alternate route remembered from his single decker bus driving days. Fortunately, there were no passengers on the upper deck that day. Why? Because that upper deck was sheared off as the bus went under an overpass too low for anything higher than a single-decker bus. The driver's old procedural memory had surreptitiously crept back from his single-level driving experiences.

Another disastrous override of conscious working memory by procedural memory occurs in what is referred to as the Forgotten Baby Syndrome (FBS). Typically, a parent or other caregiver forgets to remove a child under his/her care from a car. A few years ago in Washington, D.C., the mother of a preschooler was prevented from dropping her daughter off at the daycare center because of a work emergency. The child's father, who had never previously driven to the daycare center, offered to drop off their daughter on his way to the office. After securing the child in the backseat, he left from the house on a hot summer day. While driving he became immersed, as he customarily did, in the work commute, grumbling to himself about the heavy traffic while listening to his favorite talk show program.

On this fateful day, he lost memory of his daughter sitting in the back seat. Upon arrival, he parked in his assigned parking place, grabbed his briefcase from the front seat, and dashed into the office building. Several hours later, he recalled with horror that his daughter was still in the car. Sadly it was too late; his daughter was dead. How could such a tragic event occur?

In this example, the basal ganglia and other sub-cortical networks overcame the activity of the frontal lobe areas responsible for the planning and execution of future action. During the drive to the office the habit-based functioning of the basal ganglia and the amygdala overrode the frontal lobe's planned actions (dropping the child at the daycare center). Despite the father's best intentions, he remained immersed in the overly practiced pattern of driving himself to work in the most efficient manner (procedural memory).

The take-home message from the Forgotten Baby Syndrome and the bus driver's near catastrophic experience is that you should be wary whenever you are deviating from your usual routine. At such times monitor yourself, lest your procedural memory routines take over. Think of procedural memory and the habits thus formed as a default state. If you don't self-monitor, you'll do what you have always done previously. This can lead to unexpected and unnecessarily tragic disasters. Be alert to this potential memory peril.

FUTURE MEMORY

In the final pages of this chapter, I want to mention one final form of memory: the counterintuitive and seemingly oxymoronic future memory.

Ordinarily, when we speak of memory we are referring to the past. The consequences of past events or behavior on the present situation is an easy concept to wrap our minds around. As Spanish philosopher George Santayana famously put it: "Those who cannot remember the past are condemned to repeat it." No problem in understanding Santayana's point. But what about the influence of the future on the past and present? Lewis Carroll in his *Alice's Adventures in Wonderland*, had a paradoxical thought on that question: "It's a poor sort of memory that only works backwards."

Although it is easy to picture the past and present influencing the future, how can the future, which hasn't happened as yet, influence the past or the present?

Canadian entertainer Michael Bublé captures the essence of the dilemma in his song "Today Is Yesterday's Tomorrow." Not to be outdone, The Principles, a country and western group, came out with "Today Is Tomorrow's Yesterday."

George Orwell commemorated the counterintuitive notion of future memory in his increasingly prescient *1984*. Winston Smith, in his job as a forger of historical documents at the Ministry of Truth, makes up the fictional identity of a soldier, Comrade Ogilvy. Smith describes Ogilvy as a hero, killed in military action, and a model for his fellow countrymen.

"Comrade Ogilvy, who had never existed in the present, now existed in the past, and when once the active forgery was forgotten, he would exist just as authentically, and upon the same evidence, as Charlemagne or Julius Cesar." As a result, Ogilvy, who never existed in "real life," now inhabits the past, present, and future. In the creation of Ogilvy, George Orwell illustrated the linkage of past, present, and future commemorated in the creation of his memorable aphorism: "Who controls the past controls the future: who controls the present controls the past."

Neuroscience has shown that Carroll and Orwell were on to something. Brain scans suggest that every time we imagine a future possibility, we encode that imagined future into our memory. This involves the creation of a new memory, which when incorporated into the association network provides contact with the neuronal network formed during the creation of our earlier memories. The formation of the new memory is like an improv theater routine that varies in content according to time, cast, and circumstances. This variation is one of the reasons why people sharing the same experiences often remember events differently. It also goes a long way towards explaining why our memories—especially personal, emotionally nuanced memories—may sometimes be wrong.

"When you place an item in memory, it's as if you're sending a message to your future self," according to Robert Jacobs, a professor of Brain and Cognitive Sciences at the University of Rochester. "This channel has limited capacity, however, and thus it can't transmit all details of a message. Consequently, a message retrieved from memory at a later time may not be the same as the message placed into memory at the earlier time. That is why memory errors occur." Jacobs conceives of memory as a kind of communication channel which, like all communication channels, may break down.

For instance, the brain is designed to favor filling in details when only the gist of an experience can be recalled. Was the Shelby Mustang I considered buying last month outfitted with a manual or an automatic transmission? If I don't remember, it's natural to "mentally fill in the missing details with the most frequent or commonplace properties," says Jacobs.

The car must have been equipped with a manual transmission because I don't think Shelby ever made a car with an automatic transmission, I conclude, although I'm not all that sure of my memory for this fact and this car could be an exception or a conversion.

In J. G. Ballard's dystopian novel *Rushing to Paradise*, he writes of the dangers of a "collective amnesia for the future. . . . a willed refusal to face the imminent." Could this failure in future memory be part of the explanation for our response to the threat of Global Warming?

CHAPTER IV

MEMORY IN ACTION

AN ITALIAN DINNER FOR OUR SIXTEENTH PRESIDENT

Semantic memories within the brain await activation and association. If I ask you about Italian cooking, for instance, earlier established brain circuits are activated related to Italian meals you have
enjoyed in the past, restaurants visited, Italian wines, trips to Italy,
etc. Any or all of these experiences may be incorporated into your
response to my question. In preparing your response, you draw from
your repertory of Italian meals.

Obviously, the more experience you have had with Italian cuisine,
the more finally nuanced your response. If you are a true aficionado of
Italian food, each of your brain's association areas will be hyper-respon-
sive. You can literally see the veal Florentino in your visual association
cortex, smell and taste it via the olfactory bulb and its association areas.
Notice that there is no one correct response to my question about
Italian cooking to be drawn from semantic memory. It all depends on
your past experience, your knowledge of all things Italian, and your
best guess as to why I asked the question in the first place.

Episodic memories, in contrast, refer to one specific and potentially
recallable experience, such as, in this instance, your first Italian meal.
That meal might stick out in your memory as might any five-star Italian
feast you consumed over the years. As illustrated in the diagram below,
both episodic and semantic memory are part of long-term memory.

Working memory feeds into long-term memory by drawing on external and internal speech plus visual and spatial information via the central executive.

Here is a diagram illustrating all of this:

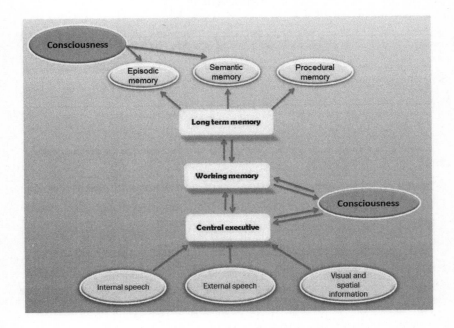

Now, for some examples of the brain in action employing different types of memory.

Learning a new language. The learning of vocabulary syntax and pronunciation involve episodic and implicit memory. Your teacher coveys the fundamentals via a series of lessons (episodes); these include words and sentences, which you learn and store in episodic memory. As you progress, you form associations for all of the things you've been taught. These are stored in semantic memory.

When you speak the language, each of your muscles of articulation, along with the muscles of your chest, mold themselves into specific shapes corresponding to the sounds of the language. If this occurs early enough in life, the throat muscles pronouncing the language coordinate their movements so that they are perfectly attuned to that language. The result is the ability to speak the language without an accent. The coordination of these motor movements comprises a procedural memory. Finally, we have the idiomatic social back and forth (gestures, loudness, etc) that accompanies all languages as they're observed and imitated based on native speakers we encounter. All of this is stored in implicit memory.

You can probably bring to mind your first few language lessons (especially if your original teacher was enthusiastic and charismatic) because these early lessons stand out in your episodic memory. As the lessons proceeded, you became less able to identify and remember the lesson when you learned a specific phrase or rule—it's all stored in semantic memory.

Once you became really fluent in the language, you no longer had to think about or monitor your speech: it became incorporated into procedural memory. But if you enter into a conversation with a master of the language you may become hesitant, clumsily flicking back and forth between semantic and procedural memory. The result is a performance far inferior to what you could accomplish if you simply did your best at your customary level of proficiency.

Answering a question is another example of the harmony and conflict that may arise among your memory systems. "Who was the sixteenth president?" Let's assume you know the answer (Abraham Lincoln). Even if you know the answer, you are unlikely to remember the specific occasion (episodic memory) when you learned that fact (unless the school burned down or some other emotionally arousing situation occurred on that day). Rather, the answer is among the many facts stored in your semantic memory. But to respond to the question, you transferred your answer from the semantic store located in the cerebral cortex to the hippocampus, the center for episodic memory. This back and forth interplay between semantic and episodic memory entails a dynamic network with the information flowing from the cortex to the hippocampus and vice versa.

After you have answered the question, the information about Abraham Lincoln will be transferred back from the hippocampus to semantic memory in the cerebral cortex, where it will reside until somebody else asks you something about Lincoln, or you have some other reason for revisiting him.

Of the two memory systems, episodic memory is the most fragile and most prone to error. Inherently we know this, especially when stating something as a fact that we are uncertain about. This may cause us to respond, depending on the circumstances, with uncertainty and expressions of anxiety (our heart may miss a beat, we may experience a sense uneasiness in our abdomen). All because we are not entirely certain of the correctness of our answer. What should we do in such circumstances? If we are uncertain about the information provided by our episodic memory, we are better off in most cases, going with our "gut" and trusting that our intuition is correct. Why? Some memories are stored inefficiently in semantic memory and do not provide an episodic memory that is entirely clear and that we can be fully confident about. But it does provide a "hunch" that one answer appears correct. So, go with it! It's a lot better than a wild guess.

A similar situation underlies the traditional advice we are given when taking multiple choice tests: don't guess. But if one of the answers "feels right," select it, and don't let any gnawing doubt cause you to change your answer.

MEMORY TRAINING UNDER WATER

As part of the attention that is required to establish a memory, think as deeply as possible about what you are trying to memorize. If it's a word, learn the word's origin; break the word down into its components; form a mental picture of the word. Mentally picturing the word brings more brain structures into play than a more superficial approach like silently spelling the word in your mind.

In one fMRI study illustrating this distinction, brain activity was higher in those areas known to be associated with memory such as the hippocampus and nearby areas of the medial temporal lobe. This suggests that those parts of the brain were working harder to form an image. This is referred to as *deep processing*, which differs from *superficial processing*. These different categories correspond to how hard the brain is working: a lot of activity in deep processing, not so much activity in superficial processing.

Forcing the brain to work harder increases the chances for later recall, as shown by fMRI investigations. In one study, volunteers were asked to look at a list of words to be memorized. On later testing, fMRI images were more active in the medial temporal lobes for those words that could later be remembered. In general, the greater the activity, the more efficient the storage and, as a result, the greater the likelihood that the subject would remember the word.

Finally, memory works best if you duplicate as closely as possible the circumstances existing at the time you try memorizing something. Background music, for instance, can influence memory. If you listen to jazz or classical selections while memorizing, you will remember better if you listen to the same type of music when recalling the memorized material. In one fascinating experiment illustrating this principle, members of a diving club learned a list of forty words either while under water or while on land. Both groups remembered more words when they were later tested under the same conditions that existed when they had originally studied the words. Next, the experimenters turned to a more everyday example, when they compared students who would learn a word list while either standing or sitting. Both groups recalled more words if they were tested while in the position they occupied when they originally studied the word list.

CATCHPHRASES

The easiest and most common way of memorizing involves coming up with a catchphrase involving all of the elements that you are trying to remember.

In medical school, I memorized the names of the twelve cranial nerves by means of a limerick: "On Old Olympic's Towering Tops

A Finn And German Vied At Hops." Remembering the first letter of each of those words corresponds to the twelve cranial nerves: olfactory, optic, oculomotor, trochlear, trigeminal, abducens, facial, auditory, glossopharyngeal, vagus, spinal accessory, and hypoglossal.

By using a rhyming technique like this, known as an acronym, the first letter of a word in the rhyme represents a different unrelated word you are trying to remember. Acronyms don't necessarily involve a rhyme, sing-song or otherwise, but can consist of a short relatively easy assembly of letters. ROY G BIV, for instance, stands for the colors of the rainbow (red, orange, yellow, green, blue, indigo, and violet).

But the most effective acronyms are rhyming ones. That's because when we rhyme or sing information, our brains learn more quickly. My guess is (nobody knows for sure) that additional brain areas are recruited when we sing or rhyme. As a result, training methods in industry or advertising or schools use a creative shorthand for conveying information—simply write down the facts or names you are trying to memorize, and use the first letters of each word to create a silly sentence.

For instance, take the planets: "My Very Eager Mother Just Sent Us Nuts" stands for Mercury, Venus, Earth, Mars, Jupiter, Saturn, Uranus and Neptune. Try coming up with your own acronym. If no other memory method existed, the use of rhyming and linking the first letters of words or lists into catchy phrases would provide you with a memory far superior to the majority of people you are likely to encounter.

All of these memory aides work because they facilitate the formation and strengthening of neuronal circuits. And each time we activate these circuits by using one or more of these memory aides, we further solidify what we have learned. Furthermore, these circuits don't exist in isolation, but are linked up with other aspects of the particular word. For instance, if you think of the word "cat," you activate circuits related to your feelings about cats, ranging from allergies to Broadway shows to a host of personal experiences involving cats. Psychologists refer to this comprising the *"unity of knowledge."*

Does that term (unity of knowledge) suggest anything to you from our survey in Chapter II about the memory beliefs held by the ancients? It reminds me of Giulio Camillo's statement, "By means of the doctrine of loci and images, we can hold in the mind and master all human concepts and all things that are in the entire world."

Whatever you call it, the unity of knowledge serves as an important measure of our mental capacity to activate widespread and interlinking memory circuits. This is the basis for the mind maps we discussed earlier as aids in retrieving memories which are temporarily inaccessible. You can think of the links in the mind maps as nodes in the brain responsible for the different associations of a word, as in the "*cat*" example above.

ALWAYS PICK THE BIGGEST SCREEN

When we aim to form mental images of the greatest clarity we are better off separating those images so they don't overlap. We know this based on research on computer displays. If you are using an iPad for instance, you will see the same images that you would see when using a desk computer with a big screen, but there is a vast difference when it comes to committing these images to memory.

On a test of pattern detection, participants improved their performance by 200 percent to 300 percent by using large displays. The small screen participants employed less sophisticated strategies, leading to more limited solutions to the problems presented to them by the experimenters. Large displays led to higher-order thinking, leading to greater ability to achieve clear better integrated insights. This was a universal finding: the bigger the display, the better the ability to detect and remember the patterns.

Bigger displays help us rapidly find the information we are looking for and store that information in memory. Smaller displays lead to a narrower visual focus and, as a result, less memory formation. To use computerese, it's as if we use more "pixels" to scan the material and connect it to memory or you can think of it as a larger display offering more space to lay out all of the images.

Multiple displays are even more efficient: 56 percent more information is recalled when presented on multiple monitors rather than on a single screen. When using multiple monitors, the participants orient their bodies towards their images on the multiple screens; rotate their torsos; turn their heads; shift their eyes. This variation in body orientation forms the basis for *proprioception*: senses send back to our brains the positions of our body in space. Proprioception also registers all bodily movements including the movements of the eyes leading to our experience of optical flow, the continuous flow of information conveyed by optic movements.

Compare this multisource stream of images with sitting immobile before a single screen. Our eyes make few or non-horizontal movements, and the rest of our body remains completely immobile for long periods of time. In the words of Annie Murphy Paul, author of *The Extended Mind*: "The use of a compact display actively *drains* our mental capacity."

The findings about computer-display size provides support for a tenet of memory expertise dating back to the Greeks. Your mental images should be *expansive*. That way you can take in more details. You can zoom in and out on those images and take note of details you would miss on a single-compact screen. When doing some of the exercises included in this book to enhance your memory, remember to use the screen with the biggest display. When doing imaginative exercises, see the objects to be remembered on a large mental screen. That way you will be able to take in details that you would miss when forming your internal "mind's eye" display.

INTERMEZZO

"A teacher who hesitates to repeat shrinks from his most important duty, and a learner who dislikes to hear the same thing twice over lacks his most essential acquisition." Sir William Gowers

Here is an overview and survey of what we have covered so far:

Encoding is the basis for all voluntarily retrievable memory. Without encoding, recall is impossible. All encoding is initially *episodic* (something happening to you, an episode); if repeated often enough, the information in memory is transferred to *semantic memory* (general knowledge); or *procedural memory* (how to do something). Working memory involves maintaining and manipulating information until it's transferred to long-term memory in one of its three categories (episodic, semantic, and procedural). Here is a diagram illustrating the interrelations. All three of these (episodic, semantic, and procedural) are part of long-term memory.

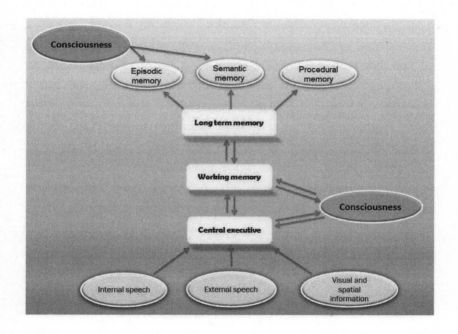

The diagram provides a spatial version of what I'm about to write. Let's start at the input, with the information intended for memory storage. This includes verbal information, both the external speech, which you hear, and the internal speech, which you recite to yourself. The other input involves visual and spatial images. Both report to the *central executive*, which monitors and manipulates these inputs. The information is then transferred to working memory.

Essentially, the central executive is *you* deciding how you will structure incoming information. All of the learning methods suggested in this book (rhyming, methods of loci, etc.) involve a *conscious decision* to use one memory method rather than another. This path between central executive and working memory travels in both directions.

Notice the pivotal role contributed by the interplay between central executive, working memory and long-term memory. The interaction of the three results in consciousness. Most important is working memory. Indeed, one could argue that "consciousness is what is occurring in working memory," according to Steve Joordens, Professor of Psychology, University of Toronto Scarborough.

This link with consciousness is most evident with episodic memory. We call it "episodic" because it consists of recalling a specific episode from your life, hence its synonym autobiographic memory. One can go even further and claim working memory is the system by which events become consciously experienced. So, memory is not only crucial to identity, but forms the basis for conscious experience. Please read that previous sentence again because it is one of the most important sentences in this book. Memory is not only crucial to identity, but forms the basis for conscious experience. So what could be a better reason for enhancing working memory?

A *feeling* of familiarity distinguishes all forms of memory. One could make a good case that the feeling of having previously encountered something or someone forms the basis for all the different forms of memory, starting with short-term (happenings that occurred just moments before); things maintained in consciousness (working memory); or things in the near or distant past (long-term memory). When this "warmth of familiarity" is missing, our memories are prone to error, with consequences we'll take up in Chapter V.

At its most basic, familiarity also forms the underpinnings for intuition. Economist and psychologist Herbert Simon stated, "intuition is nothing more and nothing less than recognition," but the recognition doesn't always operate on a conscious level.

How does human memory mature? Let's again examine again the interconnecting influences of memory, brain function, language, and behavior.

Implicit and procedural memory (so similar that for practical purposes you can group them together, if you like) are the earliest memories that develop. The infant's movement toward the breast or the bottle results from an early formed neuronal circuit—similar to but not quite a procedural memory since, as discussed earlier, the brain cannot form memories at such an early age. You can think of such actions having their origin beneath the cortex in the subcortical motor centers, which develop very early in the brain's maturation. At this point language, self-identity, and memory are not operative.

With further development we have, on each side of the brain from back to front:

1. The occipital lobe responsible for vision and pictorial memory earlier referred to as iconic memory.
2. The temporal lobe along the sides of the brain at about the level of the ears mediates sound and formulates echolalic memory.
3. The parietal lobes, just above the temporal lobes and in front of the occipital lobes, process spatial information about our bodies and the objects around us.

The development of these areas does not provide us a sense of identity or consciousness—two requirements for the formation of episodic and semantic memory. These two late developing forms of memory don't come online until the appearance of that part of the brain that distinguishes us from every other creature on earth—our frontal lobes.

Decision-making, intentional action, language, and memory (episodic and semantic) evolve in line with the development of the frontal lobes, which continue to mature well into our twenties and thirties.

All of these brain areas perform a kind of a cappella performance with episodic and semantic memory mutually influencing each other.

Think of memory as evolving in tandem with brain maturation: starting from the simplest (implicit and procedural) to the most sophisticated (episodic and semantic). The a cappella performance can't be rushed and doesn't reach its full fruition until two to five years of age, but once established lasts a lifetime.

CHAPTER V

WHAT CAN GO WRONG?

MEMORY'S MORTAL ENEMIES

When it comes to stumbling blocks that can lead to lost or distorted memories, ten examples stick out. The first seven were initially described by neuroscientist and memory specialist Daniel Schachter. These include absent-mindedness, transience, blocking, misattribution, suggestibility, bias, and persistence. Professor Schachter whimsically describes them as "sins of memory," but I will be less theological and think of them as oppositional factors limiting memory.

Absent-mindedness is the direct consequence of lack of attention. We come home from the office and discover that we are almost out of milk—the consequence of our lack of attention when we placed the almost empty container back in the fridge after breakfast. But on occasion absent-mindedness can lead to more weighty consequences than just running out of milk.

In 1999 Yo-Yo-Ma's absent-mindedness almost led to the loss of his $2.5 million cello. Hurrying out of a cab, he forgot that he had placed the instrument in the trunk. Fortunately, neither the cabdriver nor the police shared Yo-Yo-Ma's absent-mindedness and the cello was recovered and returned to him.

Transience depends on the time that has elapsed since the establishment of the original memory. Although you can recall last night's dinner, your memory of past dinners is weaker the further you delve back in time. Transience like absent-mindedness can be correlated with brain functioning. As time goes by, specific memories shift from "high" to "low" definition like a Polaroid picture. Why? Because with the passage of time neurons drop out of the circuitry responsible for that specific memory from way back in our past. You can reverse this process via conscious recollection or reminiscence. Indeed, reminiscence is used in many treatment facilities to preserve memory among older people suffering from memory problems. Typically, a specific year is selected and the participants collectively recall everything they can about that year (the president at the time, the winners of various sporting events, the important news happening around the world, etc.).

The third of the oppositional detractors from a good memory is *blocking*. When psychoanalysts speak of blocking, they refer to an emotionally traumatizing event or happening that causes sufficient distress, that the ego literally blocks it from rising to conscious awareness. What one has not consciously experienced can only be remembered with difficulty—and then only unconsciously, skirting conscious awareness. But I'm not referring to that kind of blocking. Nothing threatening or anxiety-arousing need be involved. Rather, you are trying to remember something in an unexpected framework. For instance, while shopping you encounter a colleague from work. What should be a simple recognition turns into a frenzied search in your memory "bank" for the colleague's name. You were not experiencing difficulty because you secretly dislike your colleague, as the Freudians would have it; something else was going on.

John Stuart Mill provided the common sense explanation for such situations. "Proper names are not connotative," wrote Mill, "They denote individuals who are called by them: but they do not indicate or imply any attributes as belonging to those individuals." Since every Mary in the world could just as well be named Jane or Sarah, there isn't any totally reliable way of guaranteeing that you will be able to recall the name of that unexpectedly encountered colleague in the mall. This lack of a connotative aspect to names is responsible for the most common memory failure that is complained of to doctors. Most people first encounter difficulty with remembering names rather than any other memory difficulties. This difficulty of remembering names can be improved by mnemonic techniques (see chapter 3).

More bothersome than the memory failures of absent-mindedness, transience, and blocking are "misattribution errors: Remembering something that never happened"; confusing the source of information that you correctly recall; or most disturbing of all, repeatedly remembering events that you would prefer never to think of. For instance, where were you and what were you doing when you first learned of the events of September 11, 2001 or the January 6, 2021 riot on the Capitol? Are you absolutely sure of your recall? Don't be too confident of the correctness of your memory. You may have inadvertently fallen victim to a form of *misattribution* that Schachter refers to as *source amnesia.*

Source amnesia occurs, according to Schachter, when "People recall correctly a fact they learned earlier, or recognize accurately a person or object they have seen before, but misattribute the source of their knowledge." You have probably experienced this at one time or another when you told a joke to someone and they chuckled and remarked to you that they were the person who originally told you the joke. Obviously you didn't remember them telling you this joke; if you had, you wouldn't have repeated it back to them. Many of us have difficulty distinguishing one conversation from another in terms of who said what. In psychological experiments on facial identification, participants frequently misremembered the time and place of the previous encounter with the person shown in a picture or drawing.

Three other misattribution errors are *suggestibility, bias,* and *persistence.*

We are affected by *bias* when we allow false assumptions to alter our memories. For instance, since we hold certain beliefs now it is only too easy to assume we have always held these beliefs. This consistency bias distorts our memories of the past. As an example, I personally encountered a disillusioned Trump supporter, who told a group of us over lunch that he thought many of the political problems today started with Ronald Reagan. We all laughed and reminded him that he had been a Reagan fan and that he even once proudly displayed a picture of himself taken with Reagan dating from the 1980s. My friend was revisiting his memory of Reagan and, since he no longer approved of our fortieth president, he assumed he had felt similarly in the past.

As another example, it is not unusual for memories of ourselves to serve as boosters of our ego: in our youth we were the fastest athlete, the smartest student, the most persuasive conversationalist, etc. In such instances, recollections of the past are reconstructed according to our current attitudes and perceptions of ourselves. In a phrase, the way we were is taken as a reflection of the way we are today.

In a test of this assumption, researcher Linda Levine asked a group of people how they felt after first learning about the famous acquittal in the murder trial of O. J. Simpson. To no one's surprise, the responses mirrored those found in the general population: surprise, shock, anger, and in some instances glee. But when Levine asked the same people the same question five years later, the reported feelings were not at all in line with their current feelings. If they now harbored a general feeling of resentment or anger, that's what they reported about how they had felt after the verdict. In many cases, the feelings were different, and in some cases directly opposite. What's more, most of the subjects were unaware that they had described widely different feelings five years earlier. All of this can be found in Levine's paper, "Remembering past emotions: the role of current appraisals."

Schachter mentions another study of present-induced-memory—falsification. College student volunteers were asked about the personal qualities of their current romantic partners (e.g., their honesty and intelligence). They also estimated their degree of affection for them. When asked to recall these evaluations two months later, the students' recollections tended to correspond to their current rather than their past evaluations and feelings.

The same tendency occurred in married couples that were asked similar questions. Those men and women whose feelings had changed about one another tended to mistakenly remember similar feelings in the past. While trying to remember how they had felt three years earlier, only one in five of those who were feeling differently recalled accurately how they had felt in the past. It is important to point out that this is not a matter of changed feelings on the basis of changed experiences with the romantic partners. The issue was whether or not the participants could remember how they had felt about their romantic partners at some time in the past. The majority of them failed miserably on this task, with only 20 percent expressing an accurate memory of past feelings.

Suggestibility leads to incorporating misleading information from external sources such as "authorities," written material, and the media into personal recognitions.

The seventh of Steven Schachter's "sins" is *persistence*. For some reason, our brain is better at recalling losses and failings rather than positive experiences. If you have ever performed a card trick (one of my hobbies is magic), you remember in great detail the occasions you fumbled the cards or inadvertently gave the trick away. The innumerable occasions when you successfully performed the trick is only a blur in your memory. You don't remember how many people were there, or maybe even where you performed the trick. But that one flub sticks in your memory with precision and detail: the facial expressions of the audience when they saw something they were not intended to see.

Post-Traumatic Stress Disorder (PTSD), which we will discuss more fully later in this chapter, is an exaggerated form of persistence. The memory of a car accident (the most frequent cause of civilian-based PTSD) just won't go away. You continue to experience the impact, the breaking glass, the pain in your neck and back. The more depressed you are and the more you're given to rumination (fretful worrying), the greater the persistent painful memory.

Why would persistence exert such a powerful effect on memory? Because persistence can serve as a warning sign providing a high priority to a previous event that threatened our well-being, or even our life. On other occasions, persistence may concern less pressing matters than PTSD. Sometimes we can't get a particular song or tune out of our head (so-called ear worms). Usually in such minor incidents the persistence disappears on its own when our attention is captured by something else. Persistence, especially when accompanied by depression, involves the amygdala, that almond-shaped structure located in the limbic system where emotions are processed, especially negative emotions. Psychological experiments have demonstrated that if you are shown pictures of mutilated bodies from a car crash, your amygdala will be strongly activated. A similarly strong activation will occur later, if you think back and recall those pictures. Basically, we are talking here of one of the main tenets in this book: the images that we create and entertain influence our memory, shape our reality, and help mold our personality.

In addition to the previous seven oppositional factors leading to memory failures, here are three more of my own choosing:

The eighth enemy of memory is *technological distortion*. Today, technology both helps and hinders the development of a superpower memory. Think of the technology of audio and visual recorders, computers, and electronic diaries as extensions of the brain. Thanks to these aids we can carry incredible amounts of information around with us. While this increase in readily available information is generally beneficial, there is also a downside. The storage and rapid retrieval of information from a cellphone or an iPad or a computer also exerts a stunting effect on our brain's memory capacities. We have to constantly try to overcome this by working at improving our memory. This not only will enhance our powers of recall, but will also strengthen our brain's circuits starting at the hippocampus and extending to every other part of the brain.

The ninth enemy of memory also often involves technology: *distraction*. In our day, the greatest impediment of memory is distraction. In fact, we live in an Age of Distraction. As we watch the evening news, our attention is split between what the news anchor is telling us and the contents of the crawler working its way across the bottom of our screen. Distraction prevents us from exerting the most important facilitators of memory: concentration and focus. When we concentrate, we devote all of our mental energy to one area. When we eliminate all distractions and really drill down on the area of concentration, we bring into action intense focus.

Distraction affects the working memory deficiencies that underlie poor academic achievement of adults as well as children. Attention deficit/hyperactivity disorder (ADHD) is the quintessential distraction disorder of our times. It is now so common that it's been pretty much assimilated into the realm of normally acceptable behavior. The accompanying reading difficulty frequently involves working memory problems: the earliest words in a sentence can't be recalled by the time a reader reaches the end of a sentence.

The last enemy of memory is depression. Often, the depressive feelings and expressions (sadness, laconicism, reclusiveness, weeping, etc.) overshadow and conceal the decrease in memory. The affected person states they "don't want to bothered" when asked a question that requires them to consult their memory. This is particularly noticeable during formal neuropsychological tests. Here is one of those tests.

"I'm going to give you five words. Listen to these words and repeat them out loud. After five minutes I am going to ask you to repeat them again. Ready? Here are the five words.

1. Face
2. Velvet
3. Church
4. Daisy
5. Red"

Typically, a depressed person will take longer to learn the words (will require more than one repetition) and after five minutes may only remember one or two of the words. So how do I know that this memory failure is secondary to depression? Simple. When the depression is successfully treated by medications, psychotherapy, or both, the memory proficiency returns to the level existent prior to the depression.

WHEN I FORGET, WHERE DOES MY MEMORY GO?

So far we have said a lot about memory, but nothing about forgetting. As we will later see in this chapter, unlimited memory implies consequences most of us would be unwilling to accept. But what happens within our brain's architecture when we forget?

Each nerve cell has along its outer membrane numerous spines (dendrites), which multiply with the establishment of memory and shrink or thin out with forgetting. This process takes place in all animals from flies to humans. We speak here of a molecular balance involving two separate mechanisms: one for memory, the other for forgetting. If the thinning/shrinking of dendritic spines never occurs, the animal cannot replace the old memory with a new one: the mouse that has learned to run one pathway through a maze cannot adapt and run an alternative pathway through that maze. The old memory remains and the animal cannot replace it with a new one.

Flexibility implies the ability to learn new behavior. And that depends on altering the already established circuits of the old memory and setting down circuits representing the new memory.

Forgetting is usually looked upon as a personal failure. If we try to remember an item of information and can't come up with it, we blame ourselves. Forgetfulness is especially worrying to us because of the fear that our memory failures may be the result of a degenerative brain disease like Alzheimer's. In most cases, such fears are unfounded: the occasional "senior moment" is commonly experienced by perfectly normal people as they age. Rather than a sign of mental decline, these episodes of temporary forgetfulness may be a side effect of the mountains of information that the brain has taken in and processed over the years.

According to this view, which is held by one school of neuroscientists, memory failures in an older person are at least partially the result of the increased amount of information an older person has to sift through in order to come up with a specific memory. As a result, the search takes longer and is more prone to failure with aging. Thus occasional failures of memory retrieval aren't signs of brain degeneration, but an inevitable effect of living longer. So there may be some truth in the classical put-down the older person directs to a younger counterpart, "I've already forgotten more than you'll ever know."

Neuropsychiatrist Scott Small, author of *Forgetting*, speaks of the benefit of not remembering: No matter how routinized our living, the continuous alterations to existing memories are vital for us to adapt to our rapidly shape-shifting worlds. Just as home remodeling often requires a combination of construction on top of demolition, the brain's optimal solution for behavioral flexibility turns out to be a balance between memory and active forgetting.

Forgetfulness can never be assumed to be total. It usually occurs somewhere along a continuum with the loss of details happening first. With each passing year, more of the details of a past event became unavailable to you. The event never completely disappears from your mental radar, but it gradually fades like one of those old Polaroid pictures where the colors go from sharply contrasted to faded to an almost monochromatic blur. Still, the event in the picture remained intact—it's only the vividness that had been lost.

Similarly, the transition from memory to forgetting is typically a slow and almost always an incomplete one. The memory of the photograph can almost always be sharpened by a conversation today with someone who was present during the taking of the picture. Except, with memories rather than photographs, disagreements may arise, especially over rather trivial matters.

With the passing of the years people remember things differently, no doubt secondary to other differences at the time and how they originally experienced the event in question. Such memory disagreements can sometimes be resolved by seeking the opinions of others, who were also involved in the original experience. But as mentioned later in this chapter, memories can be contaminated either by introducing false information through oversight or forms of "gas lighting." In addition, something "forgotten" can often be recovered by means of careful questioning (the basis for some cross-examination techniques in legal proceedings).

True forgetfulness means that we can't remember something, regardless of the techniques employed to aide our recall. As mentioned above, that determination can never be made with absolute certainty. We can never completely rule out the possibility that given the proper circumstances and the proper prompts, we might be able to retrieve that elusive memory. For example, a memory may return in a dream. Many dream about people and events they haven't thought about in years. Obviously, these memories were intact during that span of time (how else could you recall them?), but for one reason or another the memories hadn't emerged until they occurred spontaneously in the dream. One rule holds here; a memory is unlikely to be retrievable if it consists of information that no longer serves any purpose.

For instance, if you travel a lot, it is unlikely that you can remember your room numbers in all the different hotels you stayed in over the years. You might be able to remember some of them if given certain clues, such as, for example, being taken to the floor of the hotel where your room was located (perhaps the room was just across from the elevator and you remember sleeping poorly that night because of the noise). But if your hotel stay occurred many years ago, it's more than likely you have forgotten the room number and no clues will help you to retrieve it. Forgetfulness in such instances results from a very practical consideration: remembering over the span of many years such minutia as a hotel room number is no longer useful once you are no longer staying at the hotel. You don't remember the room number because your brain purposely never moved that information from short-term to long-term memory. And since that transition never took place, your memory for that room number never extended much beyond the time you stayed in the hotel.

The brain operates according to a "need to remember" principle. Those things that are important or might even possibly prove to be important in the future are remembered, as unimportant things are forgotten. And doesn't that make a great deal of sense?

THE MAN WHO COULD NOT FORGET

Remembering every hotel room or telephone number you have ever encountered would be more of a burden than a blessing. We know this based on a famous example described by the Russian neuropsychologist Aleksander Luria in his book *The Mind of a Mnemonist.*

One afternoon in April 1929, a newspaper reporter was referred by his editor to the Moscow Academy of Communist Education for memory testing. The reporter's memory extended beyond anything the editor had ever encountered. As with all written newspaper pieces, the reporter's stories contained many of the kinds of highly specific but trivial details that can enrich a story (names, articles of clothing, appearance, street addresses, telephone numbers, etc.). But the reporter was never observed to take notes of any kind. Despite this, his articles were filled with revealing details, and he never made a mistake.

The reporter was referred to Aleksander Luria, who is today considered the founder of neuropsychology—a branch of psychology based on a detailed knowledge of the brain.

Upon arrival at Luria's clinic, the journalist D. Shereshevskii—referred to as simply S. in Luria's notes—demonstrated memory powers beyond Luria's experience with any previous subject. S. could learn long lists of words and numbers and repeat them back hours or days later. Luria was astonished: S.'s performance never varied.

An increase in the length in the series did nothing to delay S.'s response, nor did it increase his percentage of errors. The sequences could be recalled even years later, as Luria discovered when he asked S. to return after sixteen years and tested him with a fifty-word list that he had given him those sixteen years earlier. He responded correctly and without a pause.

Luria wrote in his book *The Mind of a Mnemonist*: "I had to admit that the capacity of his memory simply had no limits." Today, *The Mind of a Mnemonist* remains a classic in describing what everyone now recognizes as a superpower—*episodic memory*, or the memory from personal experiences as opposed to *semantic memory*, which allows us to remember general facts (such as whales can be found in the ocean).

In 2017, Reed Johnson, who had majored in Russian and worked for a while in a memory research lab, tracked down a relative of S., who was living in Brooklyn. One of the insights that the relative conveyed to Johnson was that memorization did not come without effort for S., known as *The Man Who Could Not Forget*.

"For him, remembering took conscious effort and a certain creative genius. He was not a photographer, I have come to think, so much as an artist—a person who painted not from memory, but *with* memory, combining and recombining his colors to make worlds only he could see. His extraordinary case also reveals something of how our ordinary minds remember and how often they do not."

After conferring with his editor, S. decided to take up the life of an itinerant memory performer (mnemonist). Timid and shy, S. hired a circus trainer as his manager and training assistant. In order to learn how to entertain, S. was coached by a juggler.

Periodically over the years, Luria would meet with his patient. At these meetings, Luria learned that S. used three processes to enhance his memory. First, he aimed at forming rich visual images. In some cases, he could by means of a visual image remember words in other languages or even nonsense words. The key was the richness and vividness of the image.

The second process was the creation of a story appropriate to the images. The story could be fantastic or highly improbable, but followed the outline of a narrative. Third, he placed the images in familiar locations in Moscow where he lived. For instance, he favored stops on an oft-traveled street where he would place the images for later retrieval. Finally, S. was gifted with strong synesthetic powers.

At one point, S. said to Luria, "What a crumbly yellow voice you have." On another occasion, when Luria asked him how he could maintain a permanent image of a fence as a memory site ("loci," as Luria referred to them), S. spoke of a "salty taste and the fence feels so rough. Further it has such a sharp piercing sound." In short, S. was not just forming vivid images and memorizing them for translation into the words he was trying to remember, but he was employing his other senses such as taste, smell, and hearing along with synesthesia.

While S.'s superpower memory conveyed many advantages, it proved a liability in the long run. Overrun with memories, many of them mundane and useless (strings of numbers, letters, and words from many years earlier), S.'s "remarks were cluttered with details and irrelevancies," wrote Luria. In addition, despite S.'s incredible memory powers, he led a very unhappy and relatively unsuccessful life. Tortured by his memories, S. tried forgetting them by writing down his lists and burning them, but all to no avail. Finally, in despair he turned to drinking. S. died in 1958 from alcohol-related physical complications.

After a lifetime devoted to furthering his occupational achievements, S.'s memory ability overwhelmed every aspect of his mental functioning and limited his occupational accomplishments. In short, S. serves as a warning that no one should wish to remember everything that they have ever encountered. Forgetting, as it turns out, is just as important as remembering. Fortunately for our mental health, our brain discards useless information, so we can remember what is most important to us.

A SENSE OF FAMILIARITY

Have you ever felt "totally fed up" with the parade of ads on TV, or political sloganeering banners that burgeon to overwhelming numbers around election time? Of course you have. But don't expect either example to lessen anytime soon. These intrusive messages are intended to make you feel more familiar with one candidate over another. Studies have shown that a voter is more likely to vote for the political candidate who has most thoroughly saturated the airways. Of course, there are limits to this. A confirmed conservative is unlikely to vote for a candidate of the opposite persuasion, no matter how many banners or TV ads they've encountered. One thing is certain, however. If a potential voter is sufficiently exposed to the name of the politician, they will become abundantly *familiar* with the intended message:

"What the candidates hope is that their name will stand out, and that you will be drawn to them as your selection. How can they help that along? Well, by previously exposing you to their name over and over and over. They don't even care if you think much about their name, they just want that name to be a pattern you know, a pattern that has been primed and enhanced, a name that feels warmly fluent, a fluency you may indeed attribute to them being the right person for the job," according to the University of Toronto Scarborough psychology professor Steve Joordens in his excellent lectures for *The Great Courses*—"Memory and The Human Lifespan."

When we experience something repetitively, we become better—more fluent—at recognizing it. Neuropsychologists refer to this as a *perceptual fluency*. The more frequently we see a person—in varied social contexts—the easier it is to recognize them. When the occasions are a few in number or limited to just a single social setting, embarrassing errors can occur. "I didn't recognize you with your clothes on," clumsily blurted out a man when he encountered his female neighbor at a supermarket, rather than at the neighborhood swimming pool. Fortunately, the neighbor's husband, accompanying his wife, understood what was meant and laughed along.

Neuropsychologists have experimentally induced the sense of familiarity in a host of experiments. For instance, participants can be convinced that they have seen a word previously when it is contained in a list of twenty or thirty other words, if that word is more legible than the other words on the list. For reasons not fully explained, participants also tend to judge a word printed with greater clarity than others on a list as having been encountered before. In this instance, the brain correctly perceives the word as somehow different from the others, but visual clarity is incorrectly assumed to result from previous experience with the word.

If the experimental participant is asked to identify the words he "likes" along with those he "dislikes," the previously encountered word will be liked and the new words disliked—a free-form demonstration of the powers of familiarity. You can even tell the subject ahead of time that one of the words will be on the screen for a greater or lesser period of time. It doesn't matter whether you told the subject "longer" or "shorter," the previously presented word will be chosen as familiar. Since all the words appeared for the same period of time, this is called a "sham" experiment in the psychological literature. These experiments yet again remind us that our brain is wired to be most comfortable with the familiar.

Feelings of familiarity are often based on unconscious perceptions and events that we cannot consciously recall but may continue to exert powerful and unsuspected effects in our mental lives. Summarizing a considerable amount of research, I think it is fair to say that we should do away with the belief that we either can recall something from the past, or at least, we can with varying degrees of effort dredge it up; or that we have forgotten it altogether and no amount of mental spade work is going to retrieve it. Instead, it seems that we not only remember more than we know, but previous events that we cannot consciously recall may continue, due to our "unconscious memory" to exert powerful and unsuspected effects.

DO I KNOW YOU?

We can learn much about memory by familiarizing ourselves with the ways that memory can go awry. Let's start with memory kinks that can affect perfectly normal people and go on from there.

A decade or so ago, I lived in a tightly knit neighborhood where everybody knew everyone else. One of the neighbors, whose young children played with mine, seemed rather cool whenever I encountered her at neighborhood events. "Is everything ok?" I ventured after noticing the temperature had dropped even further since our last meeting. She seemed relieved that I had asked and she responded, "It would be nice if you would at least acknowledge me when we meet on the street." She then proceeded to tick off a series of encounters when passing each other. She claimed (correctly, alas) that I only spoke if greeted first and failed to register any recognition by my voice or manner. I apologized, but didn't feel confident that things would necessarily go better the next time I encountered her. Why?

I have experienced over my lifetime similar occasions when I failed to recognize or recognized too late someone familiar to me. I can trace this embarrassing facial recognition problem back several decades.

Here is another example dating a few years back when I attended a Christmas party. As my hostess brought me into the room where everyone was gathered, she said to me, "Try to spend some time talking with the judge." Since I didn't know to whom she was referring, I circulated and struck up several conversations with two or three of the guests. About an hour into the party, I found myself speaking to a pleasant-looking woman who introduced herself simply as Sandra. I realized that I had seen this woman on television and pictured in newspapers and magazines.

As our conversation flowed, I remember my hostess's suggestion and I considered asking this woman if she knew the identity of the judge. But for reasons I couldn't articulate, this conversational gambit seemed problematic. Moments later I recognized my conversant: Supreme Court Justice Sandra Day O'Connor. At what point did I remember who she was? The answer to that depends on our ideas of what it means to remember.

As I entered the room, my past experiences dealing mostly with male judges interfered with my considering the possibility that the judge mentioned by my hostess might be a woman. This unconscious and shameful sexism resulted in the creation of a false semantic category which I used when scanning the room: judges are males and no man here looks very likely to be a judge. At this point, it seems fair to say I had no awareness of the presence of Judge Sandra Day O'Connor.

Did my awareness of the true situation begin with the recognition that my conversational partner was a celebrity of some sort? Or when I remembered for the first time in an hour my hostess's remark about seeking out and talking to "the judge"? Or did it start with the ill-defined, but definite feeling of foreboding about asking, "By the way, I understand there is a judge here. Do you know who he is?" Or did awareness occur only when face and name recognition coincided via my memory of having seen her in newspaper pictures and on television?

Disturbances in facial recognition can occur at different levels in different people. It might surprise you to learn that one in every fifty people have some degree of the problem that I have been describing, *prosopagnosia* (literally, "loss of knowledge of faces"). In its fully developed form, there is a complete loss of the ability to recognize the faces of people, even though recognition occurs immediately if the person speaks, or moves, or otherwise provides some other non-facial clue. But my own experiences convince me that lesser forms of this occur in perfectly normal people. Have you ever had a similar experience?

In its fully developed form, *sufferers* from prosopagnosia (and I use the word "sufferers" advisedly since people who mistakenly conclude that they have been "snubbed" by the prosopagnostic feel devalued and resentful (like my neighbor). In response, they can often be socially vengeful.

At its worst, the person with full-blown prosopagnosia will fail to recognize even themselves or close relatives. One patient, when shown his wedding pictures, said "two people . . . one of them could be my wife because of the silhouette. . . . If it is my wife, the other person could be me." Another patient could not recognize his wife or daughters as they approached him without speaking. Another one asked his wife, "Are you . . . ? I guess you are my wife because there are no other women at home, but I want to be reassured."

Partial or attenuated forms of prosopagnosia (such as I believe I have) rarely come to the forefront simply because there are so many other ways of recognizing another person: how they walk, the sound of their voice, how they usually dress, etc. In the Sandra Day O'Connor example, since I had never previously had the opportunity to listen to her voice or observe her gait, I was completely dependent on my ability to remember her face.

Since the ability to recognize people from their faces plays such an important part in our lives, it should come as no surprise to learn that a specific part of the brain in the posterior right hemisphere (the fusiform gyrus) is involved in facial recognition. Prosopagnosia usually results from a brain lesion in this area. But my interest in this fascinating disorder has less to do with where the causative injury is located within the brain, than it does with what it tells us about memory.

If you monitor a patient with prosopagnosia, so that you can measure skin conductance (a physiological measurement) you will find that the patient, when looking at a photograph and matching names to faces, scores no better than by guessing. Yet the skin's conductance response occurs maximally when name and face match more than 60 percent of the time. In other words, overt recognition of the faces may be missing as measured by the patient's answers, but covert recognition, as measured by bodily electrical skin measurements, may occur. Lack of awareness and familiarity rather than a simple failure of recognition explains the discrepancy between what is said (I don't recognize this person) and what the body reveals through skin conductance.

As an example of the kind of difficulties that can ensue consider S. P., an amateur artist who had suffered brain damage in the relevant area in the right hemisphere. She fails to recognize familiar faces despite retained ability to identify them by voice. When asked to identify a portrait she had painted previous to her brain injury, she does well, but only by means of an elaborate scheme involving skillful deductions from the sitter's apparent age, gender, dress, and other details. She experiences these difficulties because her perceptions of the faces she had painted lack any sense of familiarity. Absent insight into her impairment, she cannot monitor and cannot correct her mistakes.

As often occurs in psychology, an extreme deviation in one direction (prosopagnosia) coexists with overdeveloped abilities (super-recognizers). It is estimated that about two percent of the population possesses superior powers of facial recognition. Think of an inverted U with the majority of people fitting into the middle of the curve, the prosopagnostics at the lower end, and the super-recognizers at the upper end of the curve.

A super-recognizer can remember a face even if only seen in profile. They can remember children's faces and show no decrease of accuracy when working with faces of different racial or ethnic origin. If pressed, they can recognize people from their eyes alone. One expert in super-recognizers compares normal facial recognition to a snapshot level representation, while super-recognizers can encode another's face, like a picture taken from many different angles.

In a "Before They Were Famous Test," pictures of movie stars were shown as adults and also when they were as young as three years of age. The super-recognizer experienced no difficulty in matching the photos.

Not surprisingly, super-recognizers have been welcomed into police and surveillance units. The London Police employed twenty super-recognizers and tasked them with identifying more than six hundred suspects thought to be responsible for riots that occurred in 2011 in London and other English cities. The super-recognizers looked at thousands of closed circuit television images.

One super-recognizer identified nearly a third of the rioters. One of the rioters identified by the super-recognizer was wearing a bandana that covered everything but his eyes. Even with such a challenging identification, the super-recognizer turned out to be right: the rioter confessed and was sentenced to prison.

How did the super recognizers compare with police facial recognition software? Police facial recognition software identified only one rioter—a rioter already identified by a super-recognizer. Although super-recognizers are not perfect—even they misidentify or fail to identify a face now and again—they consistently outperform every computer program currently in use.

Even among perfectly normal people, who are neither prosopagnostic or super-recognizers, the recognition of familiar faces can be altered by feedback from others. In a famous experiment, a psychologist convinced one of his students to stand outside a London hotel where her parents were staying. Since the parents believed their daughter was home in Australia, they responded with surprise and pleasure when they saw her. But the daughter had been instructed by her teacher as part of the experiment not to respond to her parents' greetings. Flustered, her father doubted his own powers of facial recognition and sense of familiarity. "I'm terribly sorry. I thought you were someone else," he mumbled as he slunk away.

As in this rather sadistic experiment, those with normally functioning facial recognition systems can be made to doubt the correctness of their recognition of faces—even those of close relatives and intimate friends. It's as if the brain makes a best guess estimate of another person's identity and then scans for indications of confirmation from the other person. And if that confirmation is not forthcoming, doubts, confusion, and even a sense of unreality may arise.

Failures of facial recognition are only the most dramatic example of failures involving the subjective experience of familiarity.

Normally we experience a neat alignment between objective and subjective recognition: things that we know and are familiar with will give us what nineteenth-century psychologist Edward Titchener referred to as the "glow of warmth" that we experience when encountering the familiar. Unfortunately, this warm glow can lead to errors on our part, or errors induced in us, wittingly or unwittingly, by others.

Two perceptional errors can affect our feeling of familiarity.

"I HAVE BEEN HERE BEFORE"

All of us experience at one time or another a feeling of familiarity when we encounter something or someone for the first time. From here it is only a step towards believing that what we feel in our déjà vu (French for "seen before") experience, actually took place at some time in the past. I'm not speaking here of delusions or some rare or subtle mental disturbance. Rather, it's the *feeling* that you have been somewhere before or that you have encountered earlier someone to whom you just have been introduced.

Déjà vu frequently occurs in temporal epileptics as part of the "dreamy states" they experience (recall that the temporal lobe is in direct contiguous contact with the memory centers). Fyodor Dostoevsky, who was himself an epileptic, insightfully described déjà vu in the character of Prince Myshkin taken from his novel *The Idiot*. Other dreamy states include: a lingering sense of foreboding; suspicion about the intentions of others; and "twilight states" during which one can't be certain that one is awake or asleep.

Déjà vu is a mismatch: a positive subjective recognition ("I have been here before") contradicted by the fact that you haven't been there previously. Understandably, if this subjective-objective alignment is disturbed, it can lead to feelings of strangeness and unease, as if encountering the uncanny. Déjà vu typically emerges under conditions of fatigue, overwork, stress, or too much alcohol. Statistics about the frequency of déjà vu suggest a high enough frequency in the general population to suggest that the majority of readers of this book have experienced déjà vu at some time in their lives. Have you?

WHY DOES MY COFFEE TABLE SEEM SO DIFFERENT?

Another failure of familiarity and memory involves just the opposite from déjà vu. In the much rarer *jamais vu* (French for "never seen"), memories involve events that never took place. Only about a third of the general population has ever experienced it. In its mildest form, the *jamais vu* experience consists of a mild, puzzling, and eerie feeling, that something that should seem familiar is no longer familiar.

One of my temporal lobe epileptic patients under good control (no epileptic seizures for years) reported to me that she had recently noticed that the coffee table in her living room seemed "altered." She did not deny that it was the same coffee table that she had purchased years before, but concluded there was "something different about it."

Further inquiry on my part elicited a "fascination" (her word) with the table, which she spent an inordinate amount of time staring at, moving to different locations in the room, and placing and replacing various items on it in an attempt to make it "more normal." Over a week or so, the feeling of unfamiliarity faded away—perhaps in response to a slight increase in the dose of her anticonvulsants that I ordered. Since *jamais vu* is uncommon, it may be difficult for you to conceive it. Here is an auto-experiment that can help.

Jamais vu, in contrast to déjà vu, can be voluntarily self-induced. The experience can be discomforting, but if you want to try it, here's how to do it. Select a common word, say the very word describing the object my patient chose, *table*.

Now repeat it out loud a hundred times—table, table, table, etc. While carrying out this exercise, you're likely to experience at some point the meaning of the word decreasing until it becomes a mere sound. Although you have encountered the word "table" many times during your life, you have never experienced it as just a word. In this auto-experiment, the word *table* will slowly start to seem strange and distinct from any earlier encounter with that word. That's because your repetition of the word over and over self-induces a brief *jamais vu* experience.

A similar but even stranger *jamais vu* experience can occur if you repeat your own name out loud about a hundred times. While doing that, concentrate intently on yourself as the person that name refers to. At some point, you will experience an uncomfortable feeling. You still know who you are, but you somehow experience the relationship of yourself and your name in a different way. Most people I know who have tried this little experiment reported incipient feelings of the "uncanny" or self-alienation. This is only a step away from what psychiatrists refer to as "depersonalization," a kind of *jamais vu* of oneself. This alteration of the "reality of reality" is also referred to by psychiatrists as de-realization.

If the feelings of de-realization and de-personalization are projected outward, then other people may come across to the person afflicted with *jamais vu* as strangers or imposters, who for unknown reasons are taking on the form and identity of a familiar figure.

An exaggerated *jamais vu* experience forms the basis for the Capgras delusion of the double. A person caught up in the Capgras delusion will insist, despite all reasoning to the contrary by others, that a loved one—even on occasion a pet—has been somehow mysteriously altered. Accompanying this strange and socially alienating condition is the complete loss of any sense of the familiar. It differs from ordinary *jamais vu* in that insight is lost.

Persons experiencing Capgras delusion are so certain of the correctness of their perception, that they will act out in response to it. Murders and attempted murders have been documented over the ages by people entrapped in this delusion. Medical historians have even suggested that accusations of witchcraft in past centuries may have been based on feelings of unfamiliarity experienced by some members of the community towards other members who were unmarried, lived alone, kept to themselves, or seemed "strange" in their interactions with others. Almost invariably, these were women rather than men. This near certainty that a single woman who preferred her own company must therefore be a witch overwhelmed all sense of reason and, as a consequence, the so-called "witches" were burned at the stake or otherwise murdered.

AN ATTACK IN THE PARK

Sometimes our memories can be false and we don't even know that they are. One of the most famous cases involved the internationally acclaimed developmental and cognitive psychologist Jean Piaget. Here is what Piaget claimed he could recall:

One day as a very young child, Piaget was taken by his nanny in a pram for a stroll in a nearby park. Suddenly, a man sprang from behind a bush and grabbed at Piaget and attempted to take him. The nanny fought him off and in the scuffle sustained scratches on both of her cheeks. Her screaming attracted a crowd including a nearby policeman, who searched unsuccessfully for the attacker. Piaget recalled for the rest of his life the faces of the people who gathered at the scene and expressed empathy to the nanny. He could recall the uniform of the policeman, the scratches on his nanny's face, and the exact location in the park where the assault occurred.

Years later, Piaget's father received an unexpected letter in the mail. The nanny, now long gone from the household, was returning a gold watch the family had given her at the time of the "attempted kidnapping" as an expression of their gratitude for her heroic efforts. She related in her letter about her recent conversion by the Salvation Army and her continued guilt about the fabricated incident.

Although she didn't state as much, she presumably had orchestrated the scenario (including the self-inflicted facial scratches) in order to bolster her position in the family. In one fell swoop she had progressed in status from an ordinary household employee to almost a member of the family. As part of this cozier arrangement, the details of the incident were repeatedly gone over in the Piaget household.

The adult Piaget—even though he now knew from his father about the deception—continued to speak about the event as if it had actually happened. Indeed Piaget—at one time the most famous developmental psychologist in the world—found the false memory so resistant to eradication that he ironically dubbed it "A memory of a memory, but false."

What is the most likely explanation for this strange series of events? I believe Piaget internalized the whole story based on the innumerable conversations he had heard and participated in concerning this melodramatic incident. Each of the stark details were encoded so sharply in his memory that after innumerable repetitions the incident took on its own separate reality from what Piaget *knew* had really happened.

"I therefore must have heard, as a child, the account of the story, which my parents believed and projected it into the past in the form of a visual memory," wrote Piaget.

Notice the importance of emotion in formulating Piaget's memory. No doubt frightened at many of the events carried out by Piaget's nanny (screaming, scratching her own face etc.), Piaget's still maturing brain etched these images into a memory, which would last his lifetime.

A similar memory for something that was not personally experienced occurred with the Belgian surrealist painter René Magritte. When he was thirteen, Magritte's mother Regina committed suicide by drowning herself in the nearby River Sambre.

According to Magritte, memory of this sad event included being present thirteen days later when his mother's body was retrieved from the river. Research suggests that Magritte was not among those present at the time of the removal of his mother's body, with her nightgown wrapped around her head. But that is how Magritte remembered it. This leitmotif of faces and heads enveloped in cloth coverings occurs throughout Magritte's work, especially in one of his most famous paintings "The Lovers II" from 1928.

According to Magritte authority Alex Danchev, author of *Magritte: A Life*, Magritte's "memory" of his mother with a nightgown obscuring her face became the "founding myth" behind many of his works. "Everything we see hides another thing, we always want to see what is hidden by what we see. . . . This interest can take the form of a quite intense feeling, a sort of conflict, one might say, between the visible that is hidden and the visible that is apparent," wrote Magritte.

As in the Piaget and Magritte episodes, vivid memory images can be recalled, even for events that either didn't happen or happened differently than remembered. The influence of such imagery on memory can last a lifetime.

"THE PARROT ON A BALCONY"

Do you believe that a picture may convince you of the reality of something that didn't happen? I learned the answer to that question firsthand one afternoon while I was attending a conference in Hawaii with my wife.

After dressing for dinner, we went down to the hotel restaurant. Outside the restaurant was a parrot in a cage. Since I have been fond of parrots since childhood and have owned one for more than twenty-five years, I spent time interacting with the parrot. After a few minutes of fun, we then went to dinner.

Upon the return to Washington, DC, my wife, who was a professional photographer at the time, showed me a series of pictures she had taken on our Hawaiian trip. One picture showed me standing on the balcony of our hotel room. I remember the balcony overlooking the Pacific, but I didn't recall myself standing on the balcony holding a parrot perched on my finger. As I looked at the picture I felt a sense of uneasiness. I remembered the evening quite differently from what I was seeing in the picture.

Yes, we had encountered the parrot, but not on our balcony. It had been perched outside the hotel restaurant where we had dined earlier that evening. When I mentioned this fact to my wife, she responded, "Don't you remember after dinner, we asked the manager if you could take the bird up to the room, so that I could take a few pictures. After I took the pictures, I brought the bird back to the restaurant."

I couldn't remember any of this. But how could I doubt what I saw plainly before me in the picture?

When forced to choose between my far-from-perfect memory and the picture, my brain favored the picture, hence my feeling of uneasiness. Had I consumed more wine that evening than I thought? But when I looked at my wife, I caught the shadow of a smile on her face. She then displayed two pictures in rapid successions. The first showed a parrot perched peacefully on a porch outside a restaurant; the second picture showed me standing alone on the balcony. She then showed me how she had cleverly transposed the image of the parrot to the balcony picture, thus creating a convincing portrait of an event that had never happened.

Even though I know now the balcony-parrot picture was a fake, I still hold two images clearly in my memory: the real event with me playing with the parrot outside the restaurant, and the false event of me holding it on the balcony of our hotel room. Which is true? Well the "real event" is true of course. But I don't think the false memory will ever go completely away. It is, as Piaget put it, "a memory of an event, but false."

THE WAR OF THE GHOSTS

Most of us remain blithely unaware of just how fragile memory actually is. Insight into this uncomfortable realization began with Frederick Bartlett, an early twentieth-century pioneer in memory research. He arrived at his central contribution after observing multiple versions of "Chinese Whispers," a popular children's game in the 1920s and 1930s. You probably know it as "The Broken Telephone Game."

To play it, six or eight people form a line and the first person in line whispers a made-up usually silly phrase ("The shoes don't fit my hand" or questions like "How much wood does a woodchuck chuck?") into the ear of the person second in line, who then turns to the third in line and repeats it. This continues to the end of the line, where the last person shouts out the phrase or question, so that everybody can hear it.

Almost always the phrase or question has been drastically altered and is often hilarious. While observing children playing this game (adults play it too, but the phrases are usually naughtier and more suggestive), Bartlett came up with the concept of *reconstructive memory*.

In his now classic experiment, Bartlett asked twenty of his university students to read a short passage of about three hundred words entitled *The War of the Ghosts*. He then asked them to recall it at intervals ranging from a few hours, days, weeks, months, or even years (upon a chance encounter with one of his students two years after he first tested her).

Bartlett observed that his students' pattern of remembering and subsequent forgetting followed a pattern. First, the participants shortened the story from three hundred thirty words to, on average, one hundred eighty words with the shortest version occurring after the longest time gap (two years). Second, participants confabulated (made up) details, so that unfamiliar parts of the story were converted into familiar ones.

Bartlett also observed that students remembered best when they had a clear and vivid mental picture of what they were committing to memory. To experience this for yourself, read with the intent to remember and repeat the following narrative:

"First of all you don't want blood. Smooth and quick may have to yield to smooth and slow, especially in the case of the very old and the very inexperienced. In some cases an electronic solution works best. The possibility of blood is entirely eliminated. Such a solution is more expensive initially but is cheaper in the long run."

How many of those sentences can you repeat? You should have little difficulty as long as one condition is fulfilled: you must recognize I am describing *shaving*. Otherwise, no mental picture is possible, nor any ability to remember the sentences.

Several decades ago, the research of Professor J. D. Bransford showed that if a passage was given without context people could rarely recall it. Simply supplying a title enabled subjects to understand and remember the passage.

"The procedure is actually quite simple. First you arrange the items in different groups. Of course one pile may be sufficient depending on how much there is to do. If you have to go somewhere else due to lack of facilities that is the next step: otherwise you are pretty much set."

One group was given the title before they read it while others weren't given any information at all. Only the group who had been given the title (washing clothes) before they had read the passage were able to recognize and remember sentences from the passage.

Bartlett had a term for our memories for such things as shaving and washing clothes. He called them *schemas*. We use schemas for all sorts of behavior (waiting in line in a restaurant, ordering the food, paying the check, etc.) Thanks to schemas, we don't have to think about each of the components of these performances, except on rare occasions. If this is your first time in a sushi bar, for instance, you may be perplexed about whether chopsticks are appropriate. But if you observe carefully enough, you will notice that Japanese diners eat sushi with their fingers and employ chopsticks for other items. Learning this and incorporating it alters your schema, also known as a "script" for eating in sushi bars.

Bartlett's key insight was his recognition that memory doesn't play back our experiences like a voice recorder, but reconstructs those experiences. This insight provided the basis for twentieth- and twenty-first-century applications in advertising and our court system.

MEMORY MORPHING

Marketing provides tantalizing examples of the influence of words on the formation of memories. "What consumers recall about prior products or shopping experiences will differ from their actual experience, if marketers refer to their past experiences in positive ways," according to Jared Zaltman, author of *How Customers Think*.

For instance, by placing the words "original since 1924" on the bottle, many drinkers of a soft drink (Stewart's Root Beer) recall drinking that beverage from a similar bottle years earlier as a child. But there was a problem here: the root beer had only become available as a soft drink ten years earlier. Prior to that, the beverage was served only as a fountain drink. The memory was false, based on assuming false information as true.

Other examples include those of adults, who have been made to remember some pretty outlandish things about their childhood experiences. This was based on supplying them with false information: having been rescued from drowning as a child by a life guard; surviving a vicious attack by a dog; sharing a balloon ride with an uncle. None of these things ever happened. Yet when told about the balloon ride and being shown a fabricated picture of their uncle with them in a balloon, 50 percent of the adults recalled such a trip as a child.

Memory morphing takes advantage of our memory's fluidity and impermanence. Each time we bring a memory to mind we alter the memory. Marketers know this and so do clever lawyers.

In a court room, it is very important that the direct interrogation and cross examination do not inadvertently (or purposely) introduce false information. For instance, suppose a witness is asked which of two cars failed to halt at a stop sign when the intersection was actually controlled by a crosslight. If the witness or one of the attorneys doesn't correct this misinformation, the witness and the jury as well will remember the false information instead of the truth. Forever after, the intersection will be recalled as being governed by a stop sign instead of a cross light. *Witness contamination* is the term for this.

Another example of witness contamination pertains to the verb used to describe an accident. Did one car "hit," "bump," "collide with," or "smash" another car? It makes a big difference which verb is used. Simply by changing "hit" to "smash," a witness can be induced to estimate a 10 mph difference in the speed of the first car. What's more, the witness who heard the descriptive verb "smashed" will be over 30 percent more likely to "remember" broken glass at the scene, despite the absence of broken glass anywhere in pictures taken of the accident. These examples and a large percentage of courtroom research on false memories come from the work of Elizabeth Loftus, author of *Eye Witness Testimony* and one of the world's experts on false memories and the effect of memory contamination on inducing false memories. (She was a defense witness in the 2021 trial of Ghislaine Maxwell on charges of sex trafficking).

GASLIGHTING IN BROAD DAYLIGHT

Even firmly established memories can be modified by influences occurring in the future. For example, if a person watches a movie and reaches a judgment about it (a low or high rating), his memory for that judgement may change after later reading a rave review of the movie. The person now may incorrectly remember *initially* forming a favorable opinion of the film even though on first exposure he did not like it all. What's more, participants in psychological experiments demonstrating this effect remained unaware that they were incorrectly identifying their original feelings about the film. It wasn't that they had changed their mind about the film after reading the review. Rather, their memory of how they have originally felt about the film was altered.

Using this technique known as *backward framing*, it's possible to alter an already established memory. This finding held true even when the participants specifically were asked to respond according to their original opinion and not their opinions after reading the review. "Life is a continual memory alteration experiment where memories continuingly are being shaped by new incoming information," according to memory-distortion expert Kathryn Braun-Latur, who has written extensively on the role of memory alteration in marketing.

Earlier we encountered similar memory distortions when discussing Stephen Schacter's "memory sin" of bias. Some of the examples of *memory alteration* secondary to bias can be quite stark. An unpleasant-tasting orange drink spiked with salt and vinegar was given to the participants in a memory experiment. The drink was then hyped by means of advertisements, enthusiastically claiming the drink was "refreshing." In response, some of the participants reported that they too had found the drink refreshing. Granted, an experiment such as this measures suggestibility, as well as memory. Highly suggestible people may be more prone than their less suggestible counterparts to undergo memory morphing. But in other experiments, memory alteration takes place almost universally. Let's test your susceptibility to memory morphing. Look at the following list of words for about one minute:

- Soda, heart, tooth, tart, taste, sour
- Bitter, good, sugar, candy, nice, cake
- Pie, chocolate, honey

Now after having read that list, write down all the words that you can recall. Then put the list aside for one minute. Distract yourself: pour some coffee, glance at the headlines of the newspaper. Now come back and read the following list:

- Soda, heart, tooth, tart, taste, sour
- Bitter, good, sugar, sweet, nice, cake
- Pie, chocolate, candy, honey

One of those words was not on the original list. Which word was it?

The false memory word is located later in this paragraph. Unless you are already well along in the development of your memory, you are likely to falsely recall distractors. Particularly troublesome are words that are semantically, linguistically, or conceptually similar to the words on the list. In the second list that we presented, the word *sweet* was present, but it hadn't been on the original list. All the other words were present on both lists with most of them related in some way to the concept of sweet. It was only too easy therefore to assume that sweet was included on the first list along with all of the other words which have to do with tastes.

Basically, memory morphing takes advantage of the fluidity of our memories, which I hope I have convinced you by now aren't encoded like videotapes or DVDs that we play back whenever we want to experience something from the past. Instead, memories can be altered to include different details depending on the motivation of those suggesting we remember things in ways discordant with how they really happened.

Unfortunately, we have to live with the fact that our brain is constructed in such a way as to be susceptible to mistaken perceptions and the expression of false information. But by increasing our memory power we can counteract this. More on this in a few pages when we take up memory wars and memory laws.

AMNESIA

Does it surprise you to learn that amnesia—the loss of memory—is quite common? As a neurologist, I see cases of amnesia on a daily basis. Head injury is the most frequent cause. Typically a person suffers a head injury in an automobile accident or has taken a fall (often within their own home). They may or may not be rendered unconscious and often only for a few seconds. But, depending on the severity of the injury and the extent of their concussion, they cannot remember events before the injury ranging from a few seconds to an hour or so. This is called *retrograde amnesia*—when the memory loss extends backward in time from the incident.

Others suffer *anterograde amnesia.* Typically they walk fine, speak coherently, and even ask questions about their accident or their fall. Hours later, they remember little or nothing of what they said and did immediately after the accident. They suffer from *anterograde amnesia*, which extends for a varying length of time after the head injury. In most cases the prognosis is excellent, but a significant number of accident victims retain some lacuna in their memories. Most of those with an incomplete recovery from their limited amnesia aren't particularly bothered. But others keep asking if they will remember the event. "Why would you want to remember something that was frightening and most certainly painful?" I ask them. This is not always reassuring. Something about losing a piece of their memory bothers them since, as we will further develop later in this book, memory is linked with identity.

But let's move on now to degrees of amnesia that would disturb anyone, if they retained sufficient memory to appreciate their loss.

WHAT WOULD IT BE LIKE TO LOSE YOUR MEMORY ENTIRELY?

Loss of memory or amnesia often forms a storyline for movies and television shows. In the 2002 film *The Bourne Identity*, Jason Bourne, a CIA assassin, awakens on a fishing boat with no memory of his name or past history. Bourne remained capable of learning new things and establishing new memories—it was only his past memories that had disappeared.

In real life, the most dramatic examples of the malfunctioning or non-functioning of a particular memory system come from the studies of severely brain-damaged individuals.

The first time you meet Clive Wearing, you know you are in the presence of an accomplished man. A distinguished medieval and Renaissance musicologist, an organist of virtuoso achievement, and a choral master of international renown, Wearing specialized in the music of the seventeenth, eighteenth, and nineteenth centuries. For instance, for the wedding of Prince Charles and Diana Spencer, he recreated with original instruments the Bavarian Royal Wedding dating back to February 22, 1568.

When I encountered Wearing as a member of the WNET New York team filming the series *The Brain*, Wearing gladly sat at a piano and performed several pieces for us. And then something odd occurred. His wife Deborah came into the room and Wearing stopped playing to greet her effusively. Yet, we had observed him talking to her only minutes before he had started playing. Even stranger, when he finished the piece, he came over to us and rein-troduced himself, even though we had spoken to him earlier at great length about his career.

In the early 1980s, Wearing contracted herpes viral encephalitis, a form of herpes that attacks the brain. The herpes virus greatly dam-aged the right and left temporal lobes (home of the hippocampus on each side) along with a good portion of his left frontal lobe. Damage to the hippocampus prevented him from forming new memories (hence his forgetting of his encounter with both his wife and our film crew only a short time earlier). Wearing's memory for anything happening to him never lasts more than ten seconds (anterograde amnesia), nor can he recall anything from his past (retrograde amne-sia). Since new memories cannot be formed, there are no memories to store or retrieve.

After awakening from the herpes-induced coma, Wearing found himself destined to spend his time sitting in a 12 x 12 hospital room, playing endless games of Solitaire and making repetitive entries in a notebook. The entries are always the same: "Now I'm completely awake for the first time in years." When the repetition is pointed out to him, he testily disclaims it. "I do not know who wrote that. It was not me."

I asked his wife Deborah about her husband's condition. "Clive's world now consists of a moment, with no past anchored and no future to look ahead to. It's a blinkered moment. He sees what is right in front of him, but as soon as that information hits the brain, it fades. Nothing makes an impression, nothing registers. Everything goes in perfectly well, because he has all his faculties. His intellect is virtually intact, and he perceives his world as you or I do. But as soon as he perceives it and looks away, it is gone for him. So it is a moment-to-moment consciousness, as it were, a time vacuum. And everything before that moment is completely a void. And he feels as if he is awakening afresh the whole time. He always thinks he has been awake for about two minutes."

Since Wearing is unable to encode or form episodic memory, he is left with only one avenue for learning: procedural memory. By frequent repetition, certain new procedural memories can be acquired, bypassing episodic memory entirely. For example, after numerous repetitions of a video scene without any memory or having seen the video, Wearing was able to anticipate certain upcoming parts without having any knowledge of how he learned of them. Despite his inability to recall the lion's share of his lifetime of stored musical history, Wearing, thanks to intact brain areas responsible for procedural memory, can still play complex piano and organ compositions, sight-read, and conduct choirs. Were it not for these outlets, Clive Wearing might well have done away with himself years ago.

"The saddest aspect of his condition is the impossibility of understanding what has happened to him," according to Deborah. "He can't grasp what's wrong with him because even as you are telling him something, he is forgetting the previous sentence. So he can never take in or understand what is wrong with him."

MEET HENRY GUSTAV MOLAISON

While Clive Wearing dwells on the extremes of memory dissolution, Henry Gustav Molaison—known over the years in the neuropsychological literature as simply H. M.—suffered from a surgically inflicted form of amnesia. As a child, H. M. frequently underwent intractable seizures that occurred ten or more times per day and threatened him with permanent brain damage or even death. The seizures recurred as the result of a back and forth rapid discharge from one hemisphere of the brain to the other. Neurosurgeons strongly suspected (correctly) that the discharge was ping-ponging from the hippocampus on one side of the brain to the hippocampus on the other side. As a treatment, they removed the hippocampus and the amygdala from both the right and left sides of H. M.'s brain. The operation was an easy one, as such things go, and gratifyingly successful: the seizures stopped. But new and perfidious symptoms appeared immediately after the operation.

H. M. began to demonstrate a very specific type of memory difficulty. He could no longer remember what he had for breakfast, piece together why he was in the hospital, or recognize anyone he had previously met, even if the meeting occurred only a few seconds ago. His psychologist Brenda Milner tested him daily, but each day they met, H. M. could not place her or recall any of their previous meetings. Even over the short duration of a lunch break, H. M. would return for the afternoon testing and not be able to remember Milner, or their previous morning testing.

A short time after his operation H. M.'s uncle died, leading to the expression of profound grief. Yet within a day or so, H. M. was speaking of the uncle as if he were still alive. When told again that his uncle was dead—had been for several months—he responded with renewed surprise and fresh grief. Despite these grave disturbances in establishing memory H. M. in contrast to Clive Wearing, maintained a serviceable memory for things that had happened in the past before his brain surgery. While Wearing suffered from both retrograde and anterograde amnesia, H. M.'s defect was largely restricted to anterograde memory loss: he couldn't form new memories.

H. M.'s personality was basically unchanged. Those who knew him before the surgery stated that it was the same H. M., but with severe memory problems. Whenever his attention was directed to one thing, he promptly forgot about what he had been doing only a moment previously. H. M. needed to scribble notes to himself to keep track of day to day events. As long as H. M. could fix his attention on one thing he could keep it in mind, but if he shifted from the immediate present, even for just a second, everything collapsed like a house of cards.

I have always been somewhat curious how H. M. was able to access his past memories. They were stored within the intact neocortex, particularly the areas having to do with seeing, hearing, touching, etc., but to re-experience them, it would seem necessary for the memory tracts to interact once again with the hippocampus. But in H. M.'s case, the hippocampus on each side of the brain was missing. This is a mystery, and I mention it only to remind readers that in the event of damage, the brain's components cannot simply be reconfigured like Lego pieces.

So which of these two tragic figures would you prefer to be? Clive Wearing or H. M.? Pause here and think about what your life would be like. If forced to choose, I think most people would select H. M. (I certainly would), since he retained access to events that occurred prior to his injury. Clive Wearing, in contrast, could not access anything in his past (retrograde amnesia), nor form new memories (anterograde amnesia).

SUDDEN MEMORY FAILURES

A fifty-five-year-old man on vacation at a seaside resort with his family came out of the water, approached his wife, and asked her, "What are we doing here? Who brought us here?" Over the next five minutes, he became increasingly agitated and continued repeating the same two questions. His anxiety increased and, although he recognized and responded to his family, he couldn't remember anything about the road trip to the seashore. Alarmed that he might be suffering a stroke, his family took him to the nearest emergency room. On physical examination, the doctor detected no abnormalities. But his mental evaluation revealed a dense amnesia that extended for several weeks before his dramatic presentation at the seashore, and up to and including his current visit to the hospital.

The neurologist called to the emergency room reassured the patient and his family and friends that everything would resolve over the next twelve hours. He could be so confident in the face of such a puzzling and alarming patient presentation only because he recognized the condition: Transient Global Amnesia (TGA). As predicted by the doctor, the amnesia cleared three hours later back at the hotel room. Although his family was jubilant at his improvement, the patient himself didn't share their joy since he remembered nothing about the incident, or the several days before the incident.

Transient global amnesia (TGA) is an uncommon but not rare form of amnesia. It primarily affects men between the ages of fifty and seventy. There is currently no universally accepted explanation for this acute onset of memory loss; specialists think that it may be related to a decrease in blood flow to the brain. Tests, however, are typically normal and the recovery is complete usually within six to ten hours. TGA rarely recurs. Possible triggers include stress, sexual intercourse, and as with this man, whole body exposure to cold water. What makes this condition so fascinating is what it tells us about memory.

In TGA, the affected person loses his memory within a time-frame of only a few minutes or even a few seconds. Moreover, this loss of memory for anything that happens during the amnesia episode remains permanent even though the patient eventually returns to normal. If any narrative is later employed by the patient to "explain" his strange experiences, it's not based on memories of the incident (there are none) but on what observers observed and related to him.

Transient global amnesia is the cousin of another acute or subacute loss of memory: fugue.

Several years ago I personally examined a thirty-eight-year-old farmer (Mr. T.), who without any warning left his new bride of two weeks and drove from his home in rural Maryland to Connecticut, a state he had never visited before. When he arrived in a small town he tried checking into a motel but, according to the night clerk, he was acting "suspiciously." When signing the book, Mr. T. had to look to his driver's license to come up with his name and address. He made several remarks to the fact that he wasn't sure of where he lived or his place of origin for the trip. He also was carrying no luggage. At this point, the night clerk called the local police. Within a few minutes the police arrived at the motel and took Mr. T. back to the station.

After perusing Mr. T.'s driving license, the Connecticut police made contact with the police in the small farm community near Washington, DC, where Mr. T. lived. Eventually, they spoke with Mr. T.'s wife, who frantically informed them she had no idea where her husband was. Over the next few hours, a neighbor from Washington, DC drove to the Connecticut police station and returned Mr. T. to his home. It was at that point that he was transferred to a hospital in Washington and I was asked to evaluate him.

Mr. T. expressed puzzlement at how he had made his way to Connecticut. He remembered working on the day of his disappearance, but had no memory of anything happening after his lunch break. Specifically, he had no memory of any intention to drive to Connecticut and when asked, could only give a sketchy and partially correct account of how he had proceeded from his home forty miles south of Washington to reach the small town in Connecticut where he had stopped for the night. When provided with the details provided by the night clerk and the Connecticut Police Report, he shook his head in disbelief at himself.

He was soon able to identify himself, but it took more than a day for full orientation to return. When he regained full awareness, he expressed anxiety about the cause of his amnesia. After starting psychotherapy and antidepressants, the story emerged of a reclusive loner who experienced great difficulty adjusting himself to his new marriage. He regretted that he was no longer able to spend most of his time alone.

Both forms of amnesia (TGA and fugue) represent extreme degrees of memory loss of sudden onset. In TGA, the experiencer retains a sense of personal identity: he knows who he is and recognizes his family and friends. It is only his current circumstances that elude him. But in fugue state, personal identity is temporarily lost.

Transient global amnesia and fugue state also differ when it comes to causation. Although neuroscientists aren't certain of the parts or circuits in the brain involved in TGA, most agree that the condition is brain-based. Fugue, on the other hand, is generally considered a psychologically based disorder. Mr. T. fled his home and travelled to a part of the country unknown to him, shedding along the way his memory and identity. This is an important distinction on the whole, but still leaves unanswered questions: how can a purely psychological disorder rob someone of their memory?

In both TGA and fugue state, memory is fragmented or lost altogether. In almost all cases of fugue but only rarely in TGA, stress is the overwhelming factor in causation. In response to his anxiety about his new marriage, the loss of identity accompanying Mr. T.'s fugue state conveniently freed him from his anxiety but at a price.

But as with many distinctions about memory, one can never be 100 percent certain. TGA can be a temporary response to anxiety and on occasion fugue states are not preceded by anything especially stressful. Thus, memory loss can be like a game of Jenga: one slip up in functioning, for whatever reason, and the whole memory edifice crashes (TGA). Or the process can proceed more slowly, as in fugue, where the amnesia is secondary to psychological forces. In both cases, the afflicted person has no idea what is happening to him or the reason why. The final distinction between the two involves the response of the experiencer. Typically, the person with TGA is upset and agitated; the person in the fugue state is calm and unconcerned.

STAIRWAYS TO DIFFERENT ROOMS

On occasion, serious impairments in memory can resolve themselves even when the brain is damaged. The late jazz guitarist Pat Martino went on to be an inspiring example of recovery from what appeared to be at the time an irreversible impairment.

While playing on stage in France on 1977 Martino, in his early thirties, "stopped playing and stood there for about 30 seconds," Martino wrote in his autobiography *Here and Now*. "During those moments of seizure it feels like you are falling through a black hole; it's like everything just escapes at the moment."

A diagnostic work-up revealed an arteriovenous malformation—a tangle of arteries and veins that may burst at any time, leading to death or serious impairment. This required an operation that proved successful with one notable exception: Martino lost his ability to play the guitar. In fact, he forgot that he had ever played the guitar.

"When you don't remember something, you have no idea of its existence. If I had known those two people standing by my bedside in the hospital were . . . my parents, I would've felt the feelings that went along with the events. What they went through would have been very painful for me. But it wasn't painful then because they were just strangers."

During his recovery, Martino's parents helped his memory through the use of family photographs and playing some of his own albums for him. Encouraging him to pick up the guitar was another form of therapy.

"Once I made the decision to try it, it activated inner intuitive familiarities, like a child who hasn't ridden their bicycle for many years and tries to do so again to reach a destination. There are moments of imbalance but it's subliminal, and it emerges after some mistakes and then it strengthens." Relearning the guitar required on Martino's part a complete re-acquaintance with the instrument.

By the mid-1980s, Martino had recovered to the point he could start recording again. In his autobiography he writes of how he recovered his ability to play.

"As I continue to work out things on the instrument," he wrote, "flashes of memory and muscle memory would gradually come flooding back to me—shapes on the finger board, different stairways to different rooms in the house. There are secret doorways that only you know about in the house, and you go there because it's pleasurable."

Martino's references to "different stairways in different rooms in the house" and "secret doorways that only you know about in the house" suggest that Martino may have been familiar with or intro-duced to the mnemonic method of the Memory Palace. In any case, his recovery was successful enough to earn him nominations for the Grammy Awards in 2001 ("Live at Yoshi's") and 2003 ("Think Tank").

MOOD-DEPENDENT MEMORIES

Our memory is also influenced by our moods. When we are feeling "blue," we tend to remember and dwell on sad memories from our past. This emphasis on the negative further deepens our low feelings and leads to a clinical depression. Indeed, the most painful aspects of serious depressions come from reviewing memories of past failures or inade-quacies, real or imagined. As the mood begins to lift, other memories start replacing the painful ones; memories that only a short time ago caused stress and anguish undergo a change; things are now remem-bered as not having been so terrible after all. With further improve-ment of mood, other more pleasant memories begin to emerge.

The interaction of mood and memory also comes into play in the formation of new memories. Students who consistently studied in a familiar or relaxing place of learning (a well-lit quiet room) will vary in how well they remember the materials according to their mood during a specific study session. Even though the learning envi-ronment doesn't change from one study session to another, mood changes, especially negative ones, can lead to impairments in the ability to learn and later remember.

Such examples of mood-dependent-memory illustrate how the present can influence both the past and the future: what you can remember in the past depends on your current state of mind, which, in turn, determines the future possibilities you are able to imagine for yourself.

WHY CAN YOU NEVER REMEMBER TO
LOAN ME THAT BOOK?

As we age, unless we make efforts to prevent it, a perfectly normal form of amnesia known as *prospective memory impairment* increases. This involves a failure in carrying out something that we fully intend to do, but forget to do. An example of a loss of perspective memory is the common and perfectly normal experience of leaving a room and going into another room for a specific purpose and, once there, forgetting what we had come to do. As a rule, such an occurrence is not a sign of a serious memory problem. Most likely, some distraction interrupted our train of thought when we were on our short journey from one room to another. In most instances of this frustrating experience, you can recover your memory by going back to your starting point in the first room and mentally replaying what you had been thinking about just prior to setting off to the other room.

Here is an embarrassingly common form of prospective memory failure. Have you ever had the experience of promising to do something for somebody and you keep forgetting about it again and again? You reassured your coworker in the office that you will bring the book tomorrow—"And this time, I won't forget it." But you do forget it. If you can identify with such an embarrassing experience, you may have been suffering from a defect in *prospective memory*: promising yourself to remember or do something in the future.

But the most common cause of a failure like this has nothing to with memory. It's caused by our reluctance or unwillingness to do what the other person is asking of us. So settle that issue first. Do you really want to loan the book to that person? Perhaps you fear your coworker won't return it? A "memory failure" is sometimes the alleged excuse for everything from capital murder (I don't remember firing a gun) to social no-shows ("Even though I forgot the date, I didn't want to go out with him/her anyway."). But in this case, let's assume you are perfectly willing to loan the book, but can't seem to remember to bring it to the office.

My guess is that prospective memory is so difficult because it is unique: remembering now to remember to do something in the future. I have had this problem myself on occasion and have a solution for it. But before I tell you my solution, take a pause here and ask yourself what you would do to remind yourself, after repeated failures of memory, to bring that book to the office?

This is the way I would do it. As I write this, the book I'm currently reading is *About Time: A History of Civilization in Twelve Clocks* by David Rooney. So, if I intended to bring that book to the office, I would envision in my mind's eye the front door as seen from inside my house in the form of a large clock. When I leave to start my day, the clock on the door will remind me of the book. What was your solution?

PURPLE! PURPLE!

A forty-year-old African American man is speaking: "Even though Marci is dead, I can see her on occasion as clearly as I see you now." Troy continues to relate the events that led up to his present predicament. Marci, a forty-five-year-old homeless woman, committed suicide four months earlier by entering one of the metro stations in Washington, DC, jumping down on the tracks, and assuming a kneeling position as the train approached the platform. Troy was the train operator.

Seconds after running over Marci, Troy halted the train and leaped from the cab. "I remember seeing blood all along the sides of the car." Frantic, he telephoned the dispatch station and repeated the words "Purple, purple"—a code used by operators to communicate that someone has leapt in front of a train. Minutes later, Troy felt "numb" as he responded to police questions. At one point, he turned distractedly toward the clean-up crew, busily engaged in the grisly task of removing Marci's macerated body from the tracks and washing her blood from the front and sides of the train.

Over the next several days, Troy slept poorly, often awakened by dreams of Marci's face looking up at him from the tracks. When driving his car, he repetitively experienced the fear that a pedestrian might leap from the sidewalk and throw themselves under his moving car. Shortness of breath and fears about leaving his house then followed. Usually a regular reader of various newspapers, Troy could no longer finish a paragraph or sometimes even a sentence before he drifted off to think of the "accident," as he refers to it. With the TV running in the background, even when he is not watching it, in order to keep some sense of aliveness within him, he startles and overreacts to sudden or loud sounds. But most frustrating of all are the "visitations."

"I can look away from the TV and she is there in the room staring up at me from the tracks with that same expression. Sometimes I can see her clearly; at other times I just have this feeling of her presence in the room. I know in my mind that it can't be her because she died. But there she is."

When I first spoke with Troy after he was referred to my neuropsychiatric practice, he had completely isolated himself and rarely left his apartment. His girlfriend described him as distant, unavailable, and given to "flying off the handle" over trivial matters.

"I just don't want to see anybody," Troy told me "they don't believe me when I talk about Marci. They just tell me all this is impossible. Or they want to argue with me about her. I am better off staying in my house by myself."

Troy's diagnosis is Post-Traumatic Stress Disorder (PTSD).

MEMORY THROUGH A DISTORTED MIRROR

PTSD is a disease caused by a distressing memory that cannot be voluntarily suppressed. It asserts itself either at random or, more commonly, in response to a trigger: something occurring in the environment similar to what was experienced during the original traumatic episode. The *re-experiencing symptoms* activated by the trigger include many of the things that Troy is tortured about: flashbacks (reliving the trauma over and over); bad dreams; frightening thoughts that won't go away. Troy is also experiencing *avoidance symptoms*: keeping away from events and people that remind him of the violent death of Marci; Troy no longer can enter any of the subway stations without triggering a panic attack.

Underlying these two classes of symptoms are the *arousal symptoms* that prove most stressful to Troy and the people around him: a heightened startle response; a feeling of being constantly on edge; difficulties sleeping; and angry outbursts. Accompanying these symptom complexes Troy has trouble remembering entire aspects of that fateful day. This amnesia is cruelly ironic. Troy remembers with intense clarity only one event from that day, which, otherwise, remains a blur.

Rather than a rare disorder, seven or eight out of a hundred people will experience PTSD at some point in their lives. The most publicized cause is war. Among civilians, auto accidents and accidents in general make up the large majority of PTSD patients. PTSD more than other conditions we will describe, illustrates the importance of emotions, especially negative emotions, as a driving force in the formation of memory.

PTSD, if severe enough, can transform into a serious potentially deadly condition. Violent outbreaks, mass shootings, suicide, family break-ups, alcoholism, drug overdoses—the list goes on. The most successful treatments involve, basically, modifying the disturbing memory. Improvement for a PTSD sufferer like Troy involves helping him to take advantage of the malleability of memory that we have mentioned throughout this book. Each time the traumatic experience is recalled, clarifications and amendments can be offered by the therapist to induce changes in memory of the traumatic experience. What's more, these modifications are based on specific changes in the brain's chemistry.

Each time a memory is recalled—any memory, good or bad—it is changed when we turn our attention to something else and temporarily "forget" the memory. As a result, the next time we remember the event or episode, the memory is very subtly altered. For instance, think back to your first date. You remembered less about it now than you did during those first few days after it occurred. With each passing year, it is fair to say, you remember less (unless you regularly reminisce about your first date for some reason).

This pattern of normal "forgetting" is mirrored at the molecular level by alterations in the efficiency of protein synthesis. Each time you think about your first date, you reconsolidate that memory via the synthesis of specific proteins in the memory encoding areas within your brain. If that protein synthesis is interfered with, the memory isn't successfully reconsolidated. We know this from research involving rats that have been fear-conditioned.

After conditioning for fear, that fear can be extinguished in the rat if the experimenter actively interferes with protein synthesis prior to the consolidation of the feared memory in the rat's brain. In other words, you don't give the rats sufficient time to consolidate a fearful memory if you interfere with the protein synthesis underlying memory consolidation. Can the same thing be done with humans? Can a person's memory for a traumatic event be altered? Yes, to both questions. Here's how it works:

Specific physical features distinguish the anxiety of PTSD from a normal response to trauma. At the time of the traumatic event, the hypothalamus and the amygdala activate the sympathetic nervous system, along with the adrenal glands sitting atop the kidneys. Increased amounts of cortisol, the chemical that enhances the body's ability to manage stress, are poured out by the adrenals in the person who develops PTSD.

PTSD is also marked by a greater than normal epinephrine release. This combination of higher than normal cortisol levels accompanied by higher than normal epinephrine levels may, if prolonged to the point of being habitual, constitute a biologic risk factor for PTSD.

Finally, people with PTSD have smaller hippocampal volumes, secondary to the death of hippocampal neurons thought to be caused by the increased exposure to the elevated stress hormones. And since the hippocampus serves as the initial entry point for the formation of new memories, PTSD sufferers typically complain of memory problems and show memory deficits on testing that correlates with the smaller hippocampus on each side of the brain as seen by MRI.

Experiments have confirmed the effectiveness of providing drugs known as beta blockers that work to shield the receptors that ordinarily attract the stress hormones, both in the brain and in the peripheral nervous system. In clinical trials, the use of these medications led to fewer signs of arousal when listening to taped narratives of their trauma, compared with those who listened to taped narratives but hadn't received the drug. This treatment was very similar to the rat experiments described above.

Futuristic use of stress hormone blockers could include their administration to ambulance workers heading out to the scene of a disaster. The survivors of the disaster could also be given the same drug at the scene as part of their initial treatment. The drug wouldn't completely erase memories of the trauma, but it would effectively diminish the emotionally traumatic aspects of those memories. There are even some suggestions that if the drug isn't immediately available, it might be given later and alter a person's memory for a traumatic event before it becomes completely consolidated. Such a drug de-emotionalizes the memory. Whatever horror that was experienced will still be remembered, but absent the anxiety, panic, and emotional storm that ordinarily accompanies it.

CHAPTER VI

THE PROMISES AND PERILS OF MEMORY

"IT IS THE STAR TO EVERY WANDERING BARK"

It seems that the more practical the uses for prodigious memories, the more they prove to be an advantage. For instance, Professor A. C. Aitken of the University of Edinburgh, who lived from 1895 to 1967, is still considered by some memory virtuosos as the best all-around mnemonist. He was a mathematician, mentalist, calculator, and an accomplished violinist. When Aitken memorized the first 1,000 digits of pi, he compared the experience to "hearing a Bach fugue." He arranged the digits in rows of fifty, with each row ten groups of five numbers. He then recited them in any one of several rhythms. Throughout his career, his primary method for remembering was to search out meaningful relationships linking music with learned information. This quote from Aitken captures the method:

"Musical memory can be developed to a more remarkable degree than any other, for we have a meter and a rhythm, a tune, often more than one, the harmony, the instrumental color, a particular emotion or sequence of emotion, a meaning . . . in the executant of an auditory, a rhythmic and a muscular and functional memory; and secondarily in my case a visual image of the page . . . perhaps also a human interest in the composer with whom one may identify oneself . . . and an aesthetic interest in the form of the piece."

For those of us who are not musicians, poetry can provide a similar correlation of rhythm and meaning. Novelist Elliott Holt describes this method for memorizing poetry in an essay he wrote for the *New York Times Magazine*, "Re-reading Poems." Holt picks a particular poem that appeals to him and reads it every day for a month. The goal is the connecting of words through their sounds, not just their meaning. This method enables Holt to notice new things in the text, and by the end of the month he knows the poem by heart.

You can follow Holt's lead by reading a poem aloud to appreciate the rhythm. You can do this either early in the morning (a "lark") or just before bedtime (an "owl"). Repeating the poem daily leads to what Holt refers to as "a deeper kind of attention, one that pushes past facile understanding to intimacy with the work. It's a kind of intuitive multidimensional concentration." Notice that Holt's method doesn't involve just repetition, but a focus on how rhyming and meaning interweave like a finally woven tapestry. To take advantage of this method, I would suggest you start with a poem written with a prominent rhythmic structure. I personally favor one of Shakespeare's sonnets. Just to review, a sonnet is fourteen lines long and ten syllables per line. It is composed of three four-line groupings (quatrains):

ABAB
CDCD
EFEF

And finishes with two rhyming lines (a couplet):

GG

Shakespeare's Sonnet 116, perhaps his most famous sonnet, is an excellent choice to start with. When re-reading it, emphasize the words at the end of each line:

Think of yourself striking a drum when reciting these words, with every other line rhyming.

ABAB
CDCD
EFEF
GG

Lines 2 and 4 end with love and remove, and the couplet with proved and loved—the rhyming here is less obvious but can be forced.

Sonnet 116:

Let me not to the marriage of true minds_
Admit impediments, Love is not love
Which alters when it alteration finds,
Or bends with the remover to remove:
O no! it is an ever-fixed mark,
That looks on tempests and is never shaken;
It is the star to every wandering bark,
Whose worth's unknown, although his height be taken.
Love's not Time's fool, though rosy lips and cheeks
Within his bending sickle's compass come;
Love alters not with his brief hours and weeks,
But bears it out even to the edge of doom.
If this be error and upon me proved,
I never writ, nor no man ever loved.

If your high school education took place at a certain period in our nation's history (1960s to the 1980s), you were probably required to memorize poetry selections such as Shakespeare's Sonnet 116. Sans any memory techniques, and often uninterested in the selections, you may have dreaded the whole process. Part of the problem resulted from your lack of any input about the selections. If you were free to choose as you are now, you would have learned that poetry provides an opportunity to commit to memory something that not only appeals to you, but enriches your life, as only poetry can.

One of my favorite poets is Stephen Dunn, who prior to his choice of poetry as a career entered college on a basketball scholarship and played semi-professional basketball after graduation. He then began a career in advertising. Unhappy with his choice (although he did well at it), Dunn shifted his life plans and chose poetry as a career. Writing in a plain soft-spoken style not far removed from prose, his lines are easy to read, understand, and most importantly for our purposes, memorize.

Dunn's poems reward memorization because he captured what one of his admirers described as "that most difficult magic of the ordinary."

Here are three memorizing suggestions from Dunn's poetry that I found helpful and enjoyable. The poems are *Happiness*, *Mon Semblable*, and *After Making Love*. Read each of them after searching them on your favorite search engine. Speak the lines into a recorder and then listen to your voice while reading the lines. Continue until you can repeat the lines on your own.

CLIMBING A GOLDEN MOUNTAIN

Think of memory and creativity as forming the opposing sides of a coin. The more vivid the image, the better it can be remembered since creativity flourishes best amidst clearly remembered images. Lisa Joy, the writer and director of *Reminiscence*, spoke to a *New York Times* reporter on how she imaginatively creates her scenes: "When I write I imagine the characters talking, I design the room, I block the scene in my head. I kind of transcribe the movie I'm looking at."

The real challenge for our memory (and intelligence) is to correlate things that aren't ordinarily thought of together. As philosopher David Hume wrote, "When we think about a golden mountain, we only join two consistent ideas, *gold* and *mountain* with which we are acquainted." The more outlandish the resulting image, the more likely we are to remember it. When we think of an image like an ice cream cone forming the neck and head of a portly man dressed as an acrobat, two parts of our brains are immediately activated. First, the frontal lobes alert in response to something not likely to be encountered in the real world. Second, the amygdalae, the portals along the memory pathways on both sides of the brain, come online.

As mentioned previously, the amygdala emotionally reacts with dissonance, confusion, or emotional arousal when encountering something that is so unlikely in the real world that it stands out distinctly from any of our previous experiences. Even when imagination involves things that have never been experienced (a golden mountain), it draws upon and combines elements that had been experienced separately (golden jewelry and a mountain).

The closest most people come to original, even bizarre and surreal images occurs during dreaming. The more dreamlike an experience, the more it sticks in our memory. Indeed, the memory for dreams can provide the impetus for creativity.

ATOMS ARE FLITTING BEFORE MY EYES

One of the most famous examples of the effect of dream imagery on creativity was associated with the discovery of the structure of benzene, a chemical used in printing and industry. Discovery of its chemical structure combines the elements of a detective story and a puzzle, with the solution revealed in a dream.

C_6H_6 is the chemical formula. Six carbons and six hydrogens. In the nineteenth century, the actual structure of this carbon-hydrogen arrangement baffled the brightest minds in chemistry. Its formula was long known, one carbon atom somehow associated with one hydrogen atom. But its structure stumped the chemists trying to discover it. One of the chemists, Friedrich August Kekule, a former architect, was working quite intensely but fruitlessly on trying to conceive the structure. Then, one night, he had a dream. Here is his description.

"During my stay in Ghent, Belgium, I occupied pleasant bachelor quarters in the main street. My study, however, looked onto a narrow alley way and had during the daytime little light. I was sitting there engaged in writing my textbook; but it wasn't going very well; my mind was on other things. I turned my chair towards the fireplace and sank into a doze.

"Suddenly atoms were flitting before my eyes. Smaller groups of atoms now kept mostly in the background. My mind's eye, sharpened by repeated visions of a similar sort, now distinguished larger structures of varying forms. Long rows frequently close together, all in movement, winding and turning like serpents. One of the serpents seized its own tail and the form whirled mockingly before my eyes. I came awake like a flash of lightning. This time I spent the remainder of the night working out the consequences of the hypothesis."

Here is the chemical structure of benzene surrounded by an image of a snake swallowing its tail.

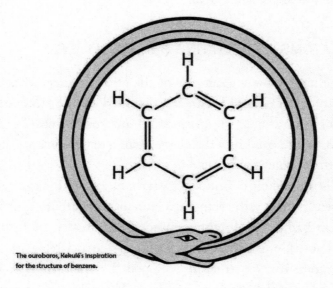

The ouroboros, Kekulé's inspiration for the structure of benzene.

Kekule was convinced that his dream originated as a recovered memory from his early studies in architecture, from which he had taken and nurtured an irresistible need for sensory experiences. He could never feel really satisfied with the explanation of a chemical phenomena unless there was visual imagery to support it. Prior to his dream, all of the elements were in place for Kekule to discover the chemical structure of benzene. All that was needed was a transformation of his chemical thinking into the distinct and bizarre image of a snake swallowing its tail. This image, an ancient symbol known as the Ouroboros, can be seen carved outside of the twelfth century Church of St. Mary and St. David in Kilpeck, Herefordshire.

Sixty years later, the connection between memory and creativity was placed on a firm neuroscientific basis, thanks to careful study of a single patient.

In the eighties, a victim of a severe motorcycle accident, K. C., was found on examination to be unable to form episodic memories. Ask him what he had been doing an hour earlier, and he had no answer for you. Much to the surprise of his doctors, he was also unable to speculate what he might be doing the following day. This was especially unusual because he lived in a unit housing patients with severe traumatic brain injury where, sadly, one day was pretty much like another. This combination of impairments in the ability to form and retain new memories, or call up in his mind's eye possible future scenes, suggested to two neuroscientists, Endel Tulving and Daniel Shachter, a connection between memory and imagination.

When hooked up to a neuroimaging machine, K. C's brain pattern of inactivation was virtually identical for episodic memory and for imagining events in the future. This suggested to Tulving and Shachter that our imagination is greatly dependent on memory; we clip and reassemble bits and pieces of past experience to model hypothetical, but as yet unrealized scenarios—in a phrase, we perform *acts of imagination*. We are talking here of the creativity that allows us to reconfigure remembered events from the past into imagined future possibilities.

Even more exciting, the interplay between memory and imagination may run in both directions. Imagine doing something in a certain way and you wind up remembering it that way.

Key to this new link between memory and imagination was the neuroscientific finding that the hippocampus is not only the initial stop on the memory pathway, but also the site where we create mental images of the world. Thus, the hippocampus plays a central role in creating mental scenes, allowing us to both experience the past and imagine the future. Brain science is thus confirming something that mnemonists have known for hundreds of years: as we progress in our ability to imagine, we strengthen our memories. That's one of the reasons why creating bizarre, dramatic, and emotionally arousing imagined scenes provide the best structure for remembering all the things that we are trying to remember.

LANDSCAPE OF A SHARED PAST

Have you ever noticed that one member of a couple often serves as the historian for their shared experiences? It's not that the other member of the couple experiences a failure of memory. It's more like he or she enjoys re-experiencing past events through their partner's fresh and different memories. "Do you remember what we did on our trip five years ago to Amsterdam?" Such a question is asked not so much because of a loss of memory for the trip, but in the hope that the response will rekindle not just a memory, but the *feeling of familiarity* that normally accompanies such recollections. This is one of the reasons the death of a spouse can sometimes be totally overwhelming. In the absence of the comfort which usually accompanies detailed and emotionally nuanced conversations with the deceased, the past seems bleak and loneliness-laden.

Tyler Wetherall wrote a marvelous essay in the *New York Times* on the pains of missing out on shared memories. When her boyfriend sustained a severe head injury, his resulting amnesia included anything having to do with Wetherall. "The span of our entire relationship had vanished." Worst of all, he could not remember the joy "and if he couldn't remember the joy, it may as well have never happened."

But Wetherall seems well aware of emotional contamination that can accompany memories, and does her best to avoid it. When speaking with him in the hospital, "I wanted an account of our story to exist independent to mine, but there was little I could do to prevent my account of our past polluting his own."

She also recognizes that the past doesn't remain frozen in time, but is as mutable as the present and the future. "In the process of telling Sam stories about our past, I had created a new story." Her overall goal was to save Sam's memories of their relationship since, "Without a partner to the collective past, these memories became less real."

Best to think of memories as both individual and shared. With shared experiences, there's always a hidden element that can only be revealed by the other person.

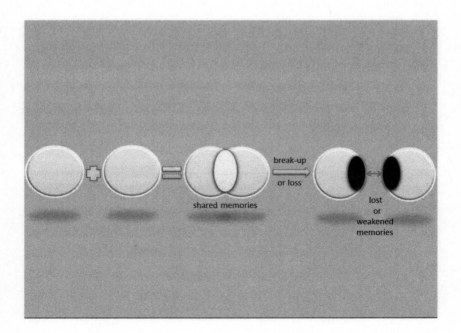

In this illustration, the overlapping hatched areas represent shared memories. Disruption of the relationship whether by "breaking up" or, as in Sam's case from brain damage and amnesia, diminishes or weakens those previously shared memories. Not only is the collective past involved, but the future is involved as well. "To break up with someone is to lose the imagined future you would create together, but you would always share your collective past," writes Wetherall.

Sadly, I can personally appreciate Wetherall's viewpoint. During the writing of this book my sister, Louise, four years younger than me, died suddenly of a cerebral hemorrhage. During our frequent get-togethers over our adult years, we enjoyed the shared experience of time traveling back to events we shared as children. Most of all, we enjoyed supplementing and correcting each other's individual memories.

Many was the time when one or the other of us forgot some aspect of our shared childhood or adolescent experiences that the other recalled. With Louise now gone, I'm left to remember events and people only through the telescope of my own memory. Did it really happen that way? Is my current interpretation of someone's past motivations accurate, or is it merely a distortion, or projection? I can no longer answer questions like these since part of my memory can't be confidently revisited nor revised. My shared past with my sister, Louise, must forever remain at the mercy of my wishful thinking and fantasy.

COLLECTIVE MEMORY

When shared memories involve more than two people, collective memories are formed and shared (such as 9/11). Like individual memories, collective memories are susceptible to alteration and amnesia. The limits of our collective memory are based on the limits of our collective experiences, interpretations, and imagination. Obviously, we can't remember what we haven't personally encountered. And clearly we can't experience events prior to our birth or before our brains matured to the point of being able to establish memories. In these instances, we rely on historical accounts along with pictures and other images like videos or movies to form a collective memory.

But as I trust we have established by now, memories are not like videos or pictures or even historical accounts. For this reason, many collective memories, especially those based on images, are subject to error. But even taking this into account, we can further distort our memories by misattributing collective cultural attitudes and values to people living in eras marked by very different attitudes and values.

Much of our culture is now locked into a mindset known as *presentism*: An uncritical adherence to present-day attitudes and beliefs, especially the tendency to interpret and *judge* past events and people in terms of current values and concepts. Presentism, like amnesia for the future, involves a loss of the ability to fully appreciate the myriad ways our current attitudes and beliefs contribute to misjudgments about the past.

The writer Gore Vidal, in mock contempt, suggested decades ago that we all are living in the "United States of Amnesia." Although Vidal's judgment seems a bit over-the-top, it leaves open this question: How do we confidently distinguish how things really were in the past from our tendency to project onto people who lived a hundred years or more ago those same feelings and thoughts that we hold today? This is one of the most important memory issues our country is currently grappling with.

A related painfully wrenching issue is how we will respond to future pandemics. Will we benefit from our unprecedented health crisis, or will we put it out of our minds—obliterate our personal and collective memories—once the immediate danger is passed? As Eric Lander, President Biden's Science Advisor, wrote in the *Washington Post* "Coronavirus vaccines can end the current pandemic, if enough people choose to protect themselves and their loved ones by getting vaccinated. But in the years to come we will still need to defend against a pandemic side effect: collective amnesia."

The government, along with social media, contribute to this collective amnesia by providing false information that can not only pollute our present perceptions, but also lead to the formation within our minds of false memories. "We live in an expository society that feeds the government our personal data and gives the tech companies the power to use it in countless other ways—and, for the most part, we expose ourselves knowingly, willingly and with all our passion and our pleasure," writes Bernard Harcourt, author of *Exposed: Desire and Disobedience in the Digital Age.*

Nor is there much chance that the surveillance society mushrooming around us is likely to diminish the distortion of our collective memories. Some of the applications are for the good, others harmful, and others yet to be determined. But no one would disagree that tracking technology ranging from surveillance cameras to smart phones is here to stay.

We are witnessing something unique in human history: everyone's public behavior and movements are the object of perpetual and intrusive observation by both public and private agents. And as a result, we have become accustomed to viewing the events and people around us in video terms. To take just one example, more and more judgments by the courts are relying primarily on video evidence. As a result, we are increasingly willing to accept as dispositive the products of such technology as surveillance cameras.

Just consider for a moment the technological intrusions that can affect you on a typical day. These include: license plate readers and cameras when you are on the highway; Google detection of smartphones via GPS satellites; Bluetooth beacons and Wi-Fi signals; Facebook live-streaming recordings; Twitter, TikTok, and Parler video captures; AT&T, T-Mobile, and Verizon logged GPS locations and Metadata; Geofence search warrants; facial recognition technology correlated with Yelp profiles; Apple iCloud account information; various instruments like Cellebrite technology, which is capable of unlocking and copying the contents of cellphones; etc.

While it is unlikely that any one of us will be subject to all of this technological surveillance, the potential is there. While a picture may be worth "a thousand words," it may also result in grievous distortions. As with my parrot-on-the-balcony experience, a picture can be modified to produce a "deep fake," depicting us in situations that never occurred. Since our memories are fragile, they can be modified by influences exerted by technological devices and other people.

Once accepted as a valid depiction of the past, any technologically defined record of events will influence our memories. We face the challenge of distinguishing what really happened (and *remembering it*) from the "explanations" and "interpretations" provided by the knotty conundrum created by the linkage of our technology and our susceptibility to presentism.

MEMORY WARS

Historical memory is the term used for the ways different people create and then identify with a specific narrative about a past event or events. Alternative terms include *collective memory* or *social memory*. Historical memories of our collective past are as fluid and susceptible as individual memories when it comes to reshaping in response to current historical-political moments. This has resulted in what psychologists and historians are describing as the *memory wars*.

For example, this sentence is being written on September 6, 2021, five days before the twentieth anniversary of September 11, 2001. So much of what we are remembering now of that momentous day depends on how old we were and where we were when we first learned of the planes crashing into the World Trade Center towers.

If we were very young, our collective memory is only a haze and borrows heavily from what we heard at the time from people who were older than us. But even among those of us who were adults at the time, our collective memory draws heavily on what we now think and feel—in the light of such things as our subsequent twenty-year retaliatory operations against the Taliban in Afghanistan.

What seemed in 2001 as a proper—indeed necessary—response to the worst attack in history on our native shores, now some parts of our culture seem moored in doubt in the face of what many see as our eventual hasty withdrawal.

In 2002, according to a *Washington Post*-ABC News poll, 55 percent of Americans said the country had changed for the better in response to September 11, 2001. Ten years later in 2011, we were roughly divided on this question. Today, nearly half (46 percent) believe the events of 9/11 changed our country for the worst, while only 33 percent say September 11, 2001 changed the country for the better. Some commentators, like *New York Times* columnist James Poniewozik, attempt to provide an explanation for this alteration in our historical memory, "The attacks set off a chain of action and changes— military quagmires, suspicion and racism at home, the loss of trust in institutions—that demagogues used to undermine democracy, and that fulfilled Osama Bin Laden's goal of dividing and weakening America."

Washington Post editor Carlos Lozada goes even further in an essay entitled "9/11 was a test. We failed."

"Rather than exemplify the nation's highest values, the official response to 9/11 unleashed some of its worst qualities: deception, brutality, arrogance, ignorance, delusion, overreach and carelessness," Lozada wrote.

Serge Schmemann, the journalist who wrote the lead article on 9/11 for the *New York Times* in 2001, has undergone a similar change of heart. "September 11th is shorthand for the moment when America lost its way, especially with the war in Afghanistan, having come to a tragic, ugly and senseless end."

Disagreements about how past events should be interpreted and incorporated into our current memory become further complicated by misinformation campaigns aimed at modifying perceptions of present events and thereby affect how these events will be remembered.

Some light was shed on this in November 2021 when the Aspen's Institute's Commission on Information Disorder came out with its eighty-page report: "Today misinformation and disinformation have become a forceful multiplier for exacerbating our worst problems as a society. Hundreds of millions of people pay the price, every single day, for a world disordered by lies," according to the report.

Disinformation is defined as false or misleading information, intentionally created or strategically amplified to mislead the public for a purpose (e.g., political, financial, or social). *Misinformation* is false or misleading information that is not necessarily intentional. But whether we are talking about disinformation or misinformation, the reality is the same: "From public health to election fraud to gender violence and predatory advertising, misinformation and disinformation result in real-world harms that impact people's lives."

Perhaps the greatest harm caused by disinformation and misinformation comes from their corrosive effects on individual and collective memory. The falsifications that we can be induced to accept and believe today will form the memories that we will be recalling in the future. What's worse, the practitioners of misinformation and disinformation aren't so much aiming at converting people from one persuasion to another, as they are at providing false information that further bolsters their already held deeply seated convictions.

As stated in the final report of the Commission on Information Disorder, "One of the most challenging aspects of addressing information disorder is confronting the reality that 'disinformation' and misinformation campaigns by bad actors don't magically create bigotry, misogyny, racism, or intolerance—instead, such efforts are often about giving readers and consumers permission to believe things that they are already predisposed to believe." Even more worrying is the suggestion by some democratic leaders that misinformation and disinformation techniques should be practiced in response to the disinformation and misinformation campaigns conducted by our adversaries.

"False, manipulated, or subverted information is a weapon," according to French Minister of the Armed Forces Florence Parly, speaking in October 2021. No one questions that providing false information is a weapon, but is it a weapon that should be employed by countries like France and, by implication, the United States?

Fighting falsehood with falsehood adversely affects not only current interpretations of world affairs, but more to our point, it affects the ways these false perceptions will influence how we will remember past events. If this sounds like the description of a war, welcome to the world of *memory wars* and their inevitable sequel, memory laws.

MEMORY LAWS

The term "loi mémorielle" (memory law) can be originally traced to December 2005, when Francoise Chandernagor wrote a highly critical article in *Le Monde* about "forcing on historians the lens in which to consider the past."

Two years later in 2007, Spain established the Law of Historical Memory directed at Francisco Franco and the Franco Regime. It aimed at not only condemning Franco, but forbidding any "exultation" of leaders or symbols related to the military coup and the Civil War of the 1930s.

So-called memory laws have been defined as: "legal provisions governing the interpretation of a historical event and the legislator's or judicial preference for a certain narrative about the past. In the process, competing interpretations may be downplayed, sidelined or even prohibited."

In some parts of the world, challenging official accounts of the past and what should and shouldn't be remembered is to bring oneself into conflict with authorities as a result of the prohibition against any expression of historical narratives that deviate or challenge the official accounts and interpretations of the past. A person who remembers things differently may be scapegoated or even subject to punitive measures.

As a result of historical revisions, events or people can even disappear into oblivion. In September 2021, Chinese movie and television actress Zhao Wei socially disappeared for several months in China. All mention of her ceased and, during her temporary banishment, her name no longer occurred anywhere on the internet—as if she had never existed.

Nor are the signs encouraging that memory laws will disappear any time soon. On this very day in 2021 that I'm writing these words, a high-level meeting is taking place in Beijing, which is expected to pass a "resolution" reassessing the Chinese Communist Party's 100-year history. According to a report from Chris Buckley of the *New York Times*, "While ostensibly about historical issues, the Central Committee's resolution—practically holy writ for officials—will shape China's policies and society for decades to come." What is expected to follow is "an intense indoctrination campaign that will dictate how the authorities teach China's modern history, textbooks, films, television shows and classrooms."

Within Buckley's alarming article is the Orwellian future predicted by Daniel Leese, historian at the University of Freiburg in Germany, and an expert on modern China. "The goal is to create a common framework, a common vision of past and future. If you don't unify the thinking of people in the circles of power about the past, it is very difficult to be on the same page about the future."

Alfred Wu, a professor at the National University of Singapore, agrees, "Controlling the narrative of history and using that to suppress alternative points of view has been a key element in the party politics. It also reveals Xi, after controlling everything from the military to decision-making, is now trying to go deeper and pursue the control of minds."

Memory laws are the embodiment of George Orwell's prescient insight about the way present circumstances can play a large part in our construction of memory. "Who controls the past, controls the future. Who controls the present, controls the past." Throughout our lifetimes, our memories are influenced by attitudes and beliefs anchored in the present and these attitudes and beliefs change with the years. Examples of this memory mutability are all around us.

Russia under Vladimir Putin exhibits all of the dangers to democracy of the *Power Vertical*: a concept introduced by Putin meaning "government from the top." In this top-down system, power is held by the elites and wielded in their interest, according to Alexander Baunov of the Carnegie Moscow Center. "You know the 'Power Vertical'. The state now wants to build a 'Memory Vertical' too." Mr. Baunov's statement came in response to Putin's attempt to change history with particular reference to the gulag.

Such an action is eerily reminiscent of a quote from a nineteenth-century Confucian scholar, "To destroy a country, you must first eradicate its history."

Ah, but this is all occurring in totalitarian countries, you may think; it couldn't happen here. Or could it?

On the same date as the release of the *Washington Post*-ABC News poll about 9/11, demonstrators removed a statue of confederate leader General Robert E. Lee from a main thoroughfare leading into Richmond, Virginia. This twelve-ton, twenty-one-foot bronze figure of Lee on a horse had been in place for one hundred and thirty years. "This is an important step in showing who we are and what we value as a commonwealth," Governor Ralph Northam stated on the occasion of its removal. Defenders of the monument responded with the claim that taking the monument down represents an attempt to erase or at least rewrite history. So who is right?

I bring all of this up about memory laws and historical memory because it is important to realize that our collective memory, just like our individual memory, can be subtly altered by enforced agendas. Whatever version of history you are exposed to in school and media, if accepted, will become your memory. To this extent, memory is hostage to whatever we're willing to accept as true. This is especially perilous if someone is exerting subtle pressures on us to favor one interpretation of history rather than another. Our acceptance of their interpretation will form the basis of our future memory. That's why we have to remain all the more astute at evaluating and asking ourselves how we will remember and what we wish to remember about our collective past, especially given current political and social forces aimed at altering our views and, by implication, our memories.

Second, we have to remain aware that our memories and interpretations of the past shape us today and in our future. As Archbishop Salvatore J. Cordileone, the archbishop of the Roman Catholic Archdiocese of San Francisco, phrased it in a *Wall Street Journal* Op-Ed, "How we choose to remember the past shapes the people we hope to be in the future."

At the moment, the most influential vehicle for creating, destroying, and distorting memories is technology.

THE LATE (?) WHITNEY HOUSTON

At its best, technology provides conveniently available applications for improving our memory. Recent photographic advances in our Apple and Android phones make possible photographs of breathtaking verisimilitude. But substituting technological memory for human biological memory comes at a cost.

Look around on your next vacation and you'll easily identify people snapping pictures at a frenzied pace. Their memories for the vacation sites will be largely limited to those pictures since little time is usually taken to really look at the subjects the camera is focused on. As a result, the pictures will not serve as prompts to personal memory, but replacements for it.

While technology can both amplify and enrich those experiences that will form the basis for our memory, it can also cause our powers of memory to atrophy. Why bother to focus, concentrate, and apply effort to visualize something when a cellphone camera can do all the work for you?

One of the ways to avoid all this is to use photographs as memory boosters to show things that you failed to perceive when you took the picture. Since it is unlikely that you will remember all of the details of your ex-college roommate's wedding dress, it's alright, for instance, to take a picture that can serve as a memory enhancer and corrector.

Current and upcoming advances in technology are already scrambling our concepts of past, present, and future. Imagine yourself watching a performance by your favorite female vocalist. You are enjoying it but for some reason you are feeling creeped out. The publicity surrounding the performance describes it as "live," but how can it be live if that vocalist has been dead for nine years?

Until recently, it was hard to argue about the finality (at least on this earth) of death. When someone died, they became a permanent denizen of the past. But now, thanks to virtual technology, we can experience a dead person in the present as if they were still alive.

In the fall of 2021, Whitney Houston, a multiple Grammy Award–winning singer, put on a sold-out performance in Las Vegas. Although the accompanying musicians and dancers were indisputably alive, "Whitney Houston" consisted of a computer-generated face of Houston at the height of her career digitally fused to an actress body-double. This high-tech afterlife performance included "Whitney" pacing the stage, calling out to her fans, and dancing so convincingly that she comes across as a real-life performer in the right now rather than someone who has been dead for almost a decade.

So how will this high-tech performance be remembered? Certainly not as something that "really happened." Rather, Houston's afterlife performance will be easily assimilated into the rapidly evolving metaverse: an alternative virtual world. And the closer the approximations of something like Whitney Houston's metaverse appearance to a live event, the harder it will be to distinguish in memory between what happened in "real life" and something occurring in the virtual world. What exactly was seen and heard?

A technological transformation of a dead person into a vivid seemingly alive recreation on the screen? A recreation believable enough to arouse uncomfortable inner experiences compatible with what computer and Artificial Intelligence experts refer to as the "uncanny valley": feelings of weirdness, strangeness, or unease in response to a virtual person taking on an almost exact resemblance to a real person?

CHAPTER VII

ACCESSORY AIDS TO A BETTER MEMORY

DRUGS

I'm often asked if there are any drugs that can boost memory. The answer is, yes, but with a qualification. Amphetamines and some other drugs are already available that can enhance memory in the short term. These drugs modify the dopamine receptor and thereby facilitate the learning of facts for an examination, for example. But such chemically aided memory for learning new material is *state dependent*: you have to take the drugs again just before the exam. Further, you are likely to forget all of those facts within hours of completing the exam. And such memory stimulants do more than just temporarily boost memory. In addition, they raise blood pressure and pulse to the point on occasion of precipitating heart attacks or strokes. Or the drugs can induce panic attacks or severe anxiety episodes. In short, the use of drugs to stimulate memory is pretty iffy.

But, these serious issues aside, does the limited (very limited) success of these drugs suggest that memory-enhancing drugs may be formulated in the future without these negative potentially fatal results? I think it does. But I don't believe that state dependency can be entirely eliminated. It is well established that modifying the brain to create a memory will most likely require a similar brain modification to retrieve it.

Other approaches may prove successful in the development of a chemically inducing memory enhancer. Work with mice has already shown that direct infusions of plasma from physically active mice result in improvements in the memory of sedentary mice. They learn to run mazes more quickly and remember the paths for longer periods of time. The mechanism seems to involve substances that reduce inflammation in the mouse's brain. One of the proteins already shown to be successful in early research (clusterin) binds to the cells lining the blood vessels feeding the brain. In the mouse form of Alzheimer's these cells are inflamed, leading to decreased memory among other things.

Two caveats must be kept in mind, however. Not everything that works in mice will turn out to be equally successful in humans. Second, if the research is going to involve humans, it would be safer to isolate an active ingredient in the plasma such as clusterin and later develop a drug featuring it.

While the future for memory-enhancing drugs seems promising in the near future, at this time more reliable memory facilitators are sleep, diet, and physical activity. Let's start with sleep.

WHY SIESTAS ARE GOOD FOR YOU

Sleep improves the quantity and quality of all of the different types of memory. Recent research has found that sleep selectively benefits memories that are encoded at a lower initial strength. In other words, when you fashion images in your memory theatre those images that are less developed are selectively strengthened during naps and sleep. The weaker the memory during the waking stage, the greater the consolidation during sleep. Furthermore, the stabilization of the weakly encoded memories makes them more resilient to subsequent deterioration.

Naps too exert a positive influence on memory. Naps lasting anywhere from thirty minutes to an hour and a half have been shown to increase later recall for information encoded prior to the nap. In one study, participants were given a choice of either napping or watching a television documentary. Both groups then studied a list of paired words. Those who napped after working on the word list, showed a 21 percent better recall than those who watched the documentary. At the level of brain activity, the nappers showed higher spindle activity (a measure of greater memory encoding) along with greater hippocampal activation measured with a fMRI. The hippocampus, as you recall, is the initial brain contact point for the formation of memory.

Several studies have come to similar conclusions regarding naps as a compensation for poor nighttime sleep. When nocturnal sleep is curtailed for some reason, a mid-afternoon nap reliably boosts memory performance, especially in adolescents and young adults.

Sleep is important for synaptic function leading to strengthening of the synaptic connections related to important memories. During deep sleep, these synapses transfer information from the hippocampus along nerve pathways to the cortex. As this takes place, long-term consolidation occurs at the cortex, independent of the hippocampus. This is the basis for the transformation of short-term memory into long-term memory.

So is napping advisable for you? If so, how best to go about it? First, not everyone successfully falls asleep during the hours best suited for naps (between 01:00 p.m. and 04:00 p.m. in the afternoon to align best with the natural circadian "siesta" dip). If you fit into this category, don't despair. Think of the nap time as relaxation time and lie down in a darkened quiet room or office with no particular intention to fall asleep. After a few days of this, your brain will gradually shift into sleep mode based on similarities to nighttime sleep (darkened quiet room, lying down on a sofa or a couch, etc.).

Now once you have fallen asleep, how do you feel when you wake up? If you feel groggy or sleep-deprived, you slept too long and the overly prolonged nap interfered with your circadian rhythm. As a result, you will have a problem getting to sleep later that night. Over the years, I've learned to sleep exactly a half hour. Some people I know have trained themselves to nap only for fifteen minutes and then awaken refreshed and reinvigorated.

The key to successful napping is waking up more empowered than you felt before the nap. In my experience, you can establish the length (from fifteen minutes to no more than an hour) and the timing (the closer to lunch, the better in order to take advantage of the natural drowsiness you feel after eating, particularly a carbohydrate-heavy meal). Although it initially takes a while to establish the nap habit, persevere with your efforts. A nap provides an incredible recharge and supercharge of your memory.

I highly recommend it.

A FEAST OR FAMINE

In regard to food, the news is also good but less specific. No diet has been identified as specifically beneficial for memory. Rather, the absence of dementia is used as a stand-in for memory, albeit not a perfect one. It is possible to possess poor memory abilities and yet not suffer from dementia. But, as I have noted on several occasions throughout this book, a highly functioning memory is virtually incompatible with Alzheimer's or any other form of dementia.

So with that as a caveat, let's start with a finding that will gladden the hearts of chocoholics everywhere. Dark chocolate enhances episodic memory in healthy young adults, according to research published in the journal *Nutrients* in 2020. Cocoa flavonoids are the ingredients in dark chocolate that improve memory function, probably via mechanisms such as increased cerebral blood flow.

Neuroscientists are now able to hone in on the part of the brain responsible for the improvement—the dentate gyrus (DG). With aging, the DG (a region of the hippocampus) declines in size and function. Neuroscientists compared fifty-two- to sixty-nine-year-old volunteer subjects who ate either a high or low cocoa flavonoid diet for three months. A high flavonoid-diet was found to enhance

DG function as measured by a high resolution variant of functional magnetic resonance imaging (fMRI), and cognitive testing. But healthy dieting isn't all about chocolate. Certain foods clearly lessen the chances of dementia.

Dr. Uma Naidoo, a nutritional psychiatrist at Harvard Medical School, has carried out impressive research which highlights the benefits of certain foods for good all-around brain health. Think of diet as an aide to help maintain memory, rather than boosting it. Enhancing memory function comes from exercising our memory with regular exercises such as those featured in this book.

Obviously, these exercises work better, if you possess a healthy brain. That is why Dr. Naidoo's suggestions make sense: a healthy brain promotes the maintenance of a healthy memory. For one thing, you will be less fatigued on a good diet and therefore more motivated to apply the techniques suggested in this book. Here are the top four suggested foods:

Berries:	Chock-full of antioxidants, minerals, and vitamins. Choose from an assortment of blue, black, and red berries. Each has slightly different flavonoids, and a mixture of all of them may confer additional benefit compared to limiting yourself to one.
Fermented foods:	When adding foods to a culture of microorganisms, the sugar in the food can be transformed into lactic acid that encourages the growth of helpful bacteria in the gut. These can include miso, kombucha, kefir, yogurt, and sauerkraut.
Leafy greens:	They contain folate, a B vitamin that aids neurotransmitter function. Included here are arugula, watercress,

spinach, Swiss chard, dandelion greens, and lettuce.

How best to incorporate these suggested foods into a healthy diet? A Mediterranean diet is high in vegetables, fruits, legumes, beans, nuts, cereals, grains, fish, and unsaturated fats, along with olive oil as a substitute for butter. The greatest transgressions against a Mediterranean diet are sugary drinks, most pizzas, fruit juices, sweetened breakfast cereals, fried food, pastries, cookies, cakes, French fries, and potato chips.

The MIND diet is another brain-oriented diet that is a bit more regimented than the Mediterranean diet:

Each day you eat three servings of whole grains such as quinoa, barley, buckwheat, brown rice etc; a salad free of any fattening caloric dressing; and another vegetable accompanied, if you wish, by a glass of wine. Snacks consist of nuts with an added half cup of beans every other day. Twice a week you can eat poultry and a half-cup of berries. At least once a week broiled or baked fish should be eaten.

Of the two diets, I personally favor the less regimented Mediterranean diet combined with some elements of the MIND diet. In the Mediterranean diet you can eat pretty much what you wish from the diet at times of your choosing as long as you avoid the prohibited foods. Since both of these diets provide even fewer carbohydrates than is currently recommended by government guidelines, they are typically associated with some weight loss. Servings of alcohol are allowed in both diets. Since I'm not a vegetarian, I may occasionally eat steaks and other meats. But these are grudging exceptions to the traditional Mediterranean diet.

Violations of both the Mediterranean and the MIND diets are much the same—namely, added sugars, fried foods, high glycemic load carbohydrates, and unrestricted amounts of alcohol.

Both diets have been suggested, but not proven to fend off Alzheimer's disease.

COFFEE AND TEA

Recent research suggests that both coffee and tea consumption are associated with a lower incidence of dementia. Particularly interesting is the finding that those who drink two or three cups of coffee and the same amount of tea showed the greatest reduction, according to the figures released in November 2021 by the UK Biobank. This study analyzed coffee and tea consumption as related to stroke and dementia risk. Among the 365,682 participants, those who drank two to three cups of coffee per day and two to three cups of tea per day lowered their dementia risk by 28 percent.

An Australian diet published in *Frontiers in Aging Neuroscience* found similar results for coffee (they did not measure tea). Coffee consumption of two to three cups per day was tied to slower cognitive decline and less cerebral amyloid-beta accumulation (harmful waste products that accumulate to excess levels in Alzheimer's disease) over 126 months.

ALCOHOL

Published opinions on this universally consumed substance vary from month to month, even day to day. Typical of the confusion is shown by the results of a study published by the French Institute of Health and Medical Research in the *British Medical Journal* in 2018. A cohort of 9087 people who didn't drink alcohol showed a higher risk of dementia compared to those who drank moderately. So what does one conclude from something like that? Everybody should take up moderate drinking and therefore outlive their teetotaling neighbors? Somehow that doesn't seem to make sense. While a small to moderate amount of alcohol is probably okay seven to twelve drinks per week maximum), I have great difficulty thinking of alcohol (a known albeit weak neurotoxin—killer of brain cells) as beneficial, or even just a neutral factor in the development of dementia.

I've searched the world literature on mild to moderate alcoholic consumption. And the news is not supportive of continued drinking, no matter how small the amount.

First of all, the purported benefits of light to moderate alcohol consumption—defined as one to at most two drinks daily, depending on the individual—is based on a "situational artifact," according to Christopher Labos, MD, a cardiologist and epidemiologist at Queen Elizabeth Hospital Complex in Montreal. Some of the research participants identified as drinkers of "zero alcohol" are actually former drinkers who had imbibed to the point they were encouraged to stop drinking altogether. Yet they are included in the control group as non-drinkers. It's likely that their former alcohol excesses are now coming back to haunt them in the form of liver and gastrointestinal diseases. This could explain why mild drinkers are found to be healthier than non-drinkers.

In order to avoid such "situational artifacts," the "zero alcohol" arm of any study assessing the benefits of small amounts of alcohol should take into account a person's drinking patterns earlier in life. Essentially, the alcohol abstinent group should be composed exclusively of people who for one reason or another *never* drank alcohol.

"When you correct for things like this then you realize that the protective effect of alcohol is either minimal or non-existent and that alcohol does more harm than good to our society," according to Dr. Labos.

When it comes to the effects of alcohol on memory—our concern here—the news is especially bleak. As I noted earlier about food, dementia must serve as a marker for memory impairment since no evidence exists—to my knowledge—that alcohol specifically impairs memory while leaving other cognitive processes unscathed. Rather, it exerts harmful effects on cognition in general—resulting, in the final stages, to dementia.

In a recent study of more than a million dementia cases in France, excessive alcohol intake was found to be one of the greatest risk factors, "even worse than things like high blood pressure and diabetes," the investigators concluded.

In the 2021 UK Biobank study of 25,000 participants, alcohol was determined to have no safe dosage. Moreover, the investigators concluded that alcohol led to changes linked to decreased memory and dementia.

Perhaps another explanation for the myth that low doses of alcohol confer health benefits stems from the findings that alcohol's damage varies with the age of the imbiber. During three periods of life, alcohol is especially harmful: 1) the gestational period extending from conception to birth—the reason pregnant women should not drink; 2) adolescence (fifteen to nineteen years); and 3) older adulthood (over sixty-five).

Based on the older adult risk factor, I advise all of my patients to abstain completely from alcohol at age seventy at the latest. By sixty-five years of age or older, people possess fewer neurons than they did only a few years earlier. So it makes sense to eliminate alcohol at a time in life when it's necessary to conserve as many neurons as possible.

EXERCISE

Exercise research over the past two decades has established the value of regular exercise. Initial claims that strenuous exercise was required to gain the benefits is no longer believed. In a representative study by Nathan Feder of the Federal University of Pelotas in Brazil, 82,872 volunteers were equally divided between men and women with the median age of 63.9 years. In this English Longitudinal Study of Aging conducted from 2002 to 2019, the results demonstrated that physical activity was associated with a lower risk for dementia. Those eighty years or older who engaged in moderate to high level of physical activity were at lower risk for dementia, compared with inactive adults aged fifty to sixty-nine years.

Even just a shift from sedentary non-activity (prolonged sitting, a "never walk when you can drive" attitude), to active movement (standing, climbing stairs, walking a mile daily, etc.) made a difference. According to a wryly amusing finding in the *British Medical Journal*, housework is linked to higher attention and memory scores and better sensory and motor function in older adults, independent of all other physical activity. This serves as another reminder that anything that gets one up and about and focuses attention, however briefly, will prove beneficial. At the deepest level, physical activity of any sort promotes synaptic and cognitive resilience.

POSTSCRIPT

LET ME TAKE YOU TO THE US MEMORY CHAMPIONSHIP

At various points in this book I have suggested memory challenges to increase your memory to some extent. I use the qualifier "to some extent" because as you continue on the path to memory mastery, you will notice that your improvement varies in direct proportion to your engagement. Goals will also vary according to how well your memory was functioning before you started reading this book. But whatever your memory's initial status, it's dynamic and can always be improved.

Since a small but significant portion of readers may aspire to the memory mastery of a competitive mnemonist, the following is addressed to them.

When I was asked to deliver the keynote address at the conclusion of the Seventeenth Annual US Memory Championship, then held at the Con-Edison headquarters in Lower Manhattan, I spent my time earlier in the day observing the day-long competition consisting of four preliminary exercises in the morning and three championship rounds in the afternoon. Here is an overview of what took place the year of my attendance.

Thirty-two mental athletes signed up for the event and competed in the qualifying rounds aimed at reducing the number of contestants from the initial thirty-two in the morning sessions to eight finalists in the afternoon.

During the morning session the thirty-two mental athletes sat at long tables, four to a table, with the monitor sitting at one end of the table to make certain all of the rules were kept. Some of the mental athletes increased their concentration and focus during these initial qualifying events by wearing dark glasses and/or headphones in order to eliminate distractions.

The first exercise in the morning is "memorization for names and faces." Each participant was given a pack of 117 color headshots and requested to memorize the first and last name of each of the faces depicted in the photos. They were then given the same photos without the names and in a different order. They had to correctly identify the first and last names for each photo. A point was awarded for every correctly spelled name.

The second event involves "memorization for speed numbers." A list of computer-generated numbers was presented in rows of twenty-eight digits with twenty-five rows per page. Twenty points were awarded for every full row that was completely recalled.

The third event was the "memorization of poetry." The mental athletes were shown a poem and asked to memorize it. They had to recall the poem from the beginning and write it down exactly as it was written. Points were awarded for every correctly spelled word, every incidence of a capital letter, and the punctuation marks in each line.

The final qualifying event in the morning involved "memorization of speed cards." The mental athletes memorized a freshly opened and shuffled pack of fifty-two playing cards. After memorizing the order of the cards, each participant placed the deck facedown on the table. Next, a second deck of cards was opened and shuffled. The contestant was asked to arrange the second deck in the same sequence as the previously memorized deck. The mental athlete who arranged all fifty-two cards in the shortest time wins the event.

Following the lunch break, the eight highest-scoring mental athletes from the morning squared off in three Championship determining rounds.

Championship Round 1: Words to Remember. Each of the participants was given the same list of words organized numerically in five columns with twenty words per column. This list consisted of concrete nouns, abstract nouns, adjectives, or verbs. They were then asked to recall as many of the words from the list as possible. The first two participants who either failed to recall a word or came up with an incorrect word were eliminated.

Championship Round 2: Tea Party/ Three Strikes You're Out. Six people from the audience were invited onstage and recited a narrative of personal information: their name, date of birth, phone number, home address, favorite pet, favorite three hobbies, favorite car, favorite food, etc. A few minutes, later the mental athletes were required to recall the information provided by the six people. The contestant's memory for all pieces of information had to be correct and complete: A date of birth must include the month, date, and year; the telephone number must be exact. Each contestant was allowed only two mistakes. The third mistake resulted in elimination from the contest (Three Strikes You're Out). This round continued until only three contestants remained.

Final Championship Round 3: Double Deck Of Cards was the decisive contest for the day. The three remaining mental athletes memorized two decks of playing cards, which had been arranged in exact order. Memorization started from the first card of the first pack and proceeded down to the last card in the second pack. Successive cards were identified in turn by each of the mental athletes. Any errors led to disqualification.

The year I attended, Nelson Dellis won the US Memory Championship and then went on to write his book *Remember It*, which I recommend.

ALL YOU EVER HAVE TO KNOW ABOUT IMPROVING MEMORY

If I have succeeded in my purpose in this book I have convinced you that:

1. Memory is the key to maintaining enduring mental acuity. If you use the materials I've suggested to augment and maintain memory, you are far less likely to fall prey to degenerative brain diseases like Alzheimer's and its dementia variants. A well-functioning memory virtually rules out these illnesses.

2. Memory is very much like physical attributes such as strength, endurance, and agility. Like them, a superior memory requires practice. Sustained practice is easy and even fun when you adopt any of the techniques I suggested, such as rhyming (sounds like), icons (looks like), or your personally constructed memory palace.

3. Memory remains fragile and can be influenced by both benign and malign forces: advertising (backward framing); politics (governmental attempts to memorialize past and current events in politically approved terms and frames); and presentism (judging the past exclusively through the perspective of currently ascendant points of view).

Here are twenty suggestions for memory improvement drawn from this book. Read them over from time to time in order to refresh the memory concepts and practices. The suggestions can also provide a quick "go-to" when deciding what form of memory you are calling upon (episodic, semantic, procedural, etc.).

1. Memory is a natural extension of attention. Just by attending to something, you increase your chances of remembering it.

2. Along with attention, concentration and imagination form the three key elements in establishing a memory.

3. Strengthening your memory leads to an increase in crystalized intelligence—the intelligence that isn't affected by aging.

4.   Working memory is the most important memory type. It is linked to intelligence, concentration, and achievement. Working memory exercises should be honed daily.

5.   Pay particular attention to sensory memory: the brain's initial recording of physical sensations as they impinge on your sense organs. Iconic memories (things that we see) and echoic memories (things that we hear) are the main forms of sensory memory. But do not neglect smell (olfaction) and taste.

6.   Memorization is most effective at times of maximum alertness. Are you an owl—more alert and concentrated late in the day and early evening hours? Or are you a lark—memorizing at maximum efficiency early in the morning and the first half of the day?

7.   Whenever you try to remember something, identify which memory system is activated. Is it short-term? Then it requires immediate recall, which lasts only a few seconds. Or is it working memory, which can last minutes or hours, depending on the purpose, intensity, and duration of your efforts? Does it require conscious awareness? Then it is either episodic or working memory. If it involves working memory, then it's likely to involve internal speech, as well as juggling of visual and spatial information. All of these sources contribute to working memory. Does it resist verbal descriptions? Then it is procedural or implicit memory.

8.   Even after you have learned something, your long-term memory for it will be strengthened if you repeatedly challenge yourself to recall it again and again. By rehearsing things already learned, your memory will become more deeply embedded.

9.   The brain tends to remember uncompleted or interrupted tasks better than completed ones. It's

best therefore to take a short break before you finish reviewing material to be memorized. By temporarily turning to some relaxing diversion, you will wind up with a stronger memory for the material than you would have if you plodded on without a break.

10. Determine during list-learning whether you are an auditory learner or a visual learner. Do you do better repeating back the list after having heard it, or after having read it?

11. Resort to mind maps when experiencing difficulty recalling specific facts, names, or events.

12. Pay particular attention to your physical surroundings when learning material and try as best you can to replicate these physical surroundings when recalling the material.

13. Once learned, material requiring the establishment of routines segues into physical actions (procedural memories). Take care not to interfere with procedural memory by shunting memory processing into conscious awareness.

14. The best way to reduce stress is by mentally reformulating stressful events or situations. It's that reformulation, which will form the basis for your memory of a stressful event.

15. The tools needed for memory enhancement include paper, pencil, chronograph, or watch with the second hand, and a voice recorder.

16. Your cellphone is an invaluable tool for memory enhancement. You can use it to generate random numbers of any length. You can then read them, memorize them via any of the methods I have described, and recall them. You can take photos of interesting and complex environments for immediate memorization. This ensures that memory includes non-verbal material.

You can record short passages of a book or newspaper article and compare your recollection of it with the dictated copy. Finally, when you are on the phone, record the conversation. Afterwards, write down everything you can remember about it. Then listen to the recording of the conversation, find out what you missed, and form a memory icon for that. For instance, if you forgot the mention of a trip to the gym, picture your conversant lifting a barbell made of feathers or marshmallows with the name of the gym emblazoned on the handlebar.

17. Use artworks as a background on which to visualize what you are trying to remember. My favorite is Edward Hopper's "Western Motel," painted in 1957 and now displayed at the Yale University Art Gallery. If you chose that picture, a large number of memory icons can be displayed on the flat surfaces, such as the extended windowsill, the bed with its long flat board, the distant mountains, which at their most distant, form a flat surface. Start by intently studying the details of the painting until you can literally see it in your mind's eye. Then describe it while looking away from the painting. Did you include the tiny clock on the bedside table? The gooseneck lamp? The piece of clothing on the chair at the lower right of the painting? Committing all these details for visualization is a memory exercise in itself. When you have committed them to memory Hopper's painting can serve as your own private memory theater. If Hopper doesn't appeal to you, select any representational painting that contains many objects suitable for serving as memory loci.

18. Learn to nap. Specifically, after you have studied for an exam or office presentation put aside time for a nap. Your ability to retrieve the necessary information will

improve secondary to the enhanced consolidation and encoding taking place in your brain during the nap.

19. Use memory images to overcome short inconsequential, but annoying terms, names or short phrases that you can't come up with. My wife's dog is a Schipperke. After repeated failures when walking Leah to effortlessly respond to the simple question "What breed of dog is that?" I formed the image of a small sailboat (small dog) with a burly skipper holding a huge key. Get in the habit of converting anything which you find hard to remember into a wild, bizarre or otherwise attention grabbing–image.

20. Create a memory theatre involving places you encounter regularly. As mentioned, I have as my memory theatre several locations which can be reached during a walk from my home. I have also a memory theatre involving a walk from my office. You really don't need more than two or at the most three. The important point is that you can see the loci links with photographic clarity.

Best wishes in your efforts to significantly enhance your memory. The sooner you begin applying the principles and techniques suggested in this book, the sooner you will reach that goal. The most important thing? Make memory improvement not just a fanciful wish, but a *daily activity*.

GLOSSARY

Amygdala An almond-shaped structure just in front of the hippo-campus. Its activation is responsible for the emotional components that accompany the formation or retrieval of a memory. Traditionally considered a mediator of negative emotions (anger, fear, etc.), we now think of it as associated with emotions in general. It also serves as a marker for items that are especially important or stand out from the background (i.e., a red color flashed somewhere in a series of blue colors). The amygdala is part of a circuit (the limbic circuit) which extends from one end of the brain to other and contributes to the arousal and expression of emotion. When you are angry, upset or gloriously happy your limbic system is "on fire."

Anterograde Amnesia Absence of the ability to form new episodic or new semantic memories. Older memories may be preserved.

Association Cortex Areas of the cerebral cortex that are involved in associating information, coming from all of the sensory inputs.

Associationism holds that memory depends largely on the formation of linkages ("associations") between events, sensations or ideas. Recalling one member of the associated link elicits memory of the others.

Basal Ganglia A group of brain structures beneath the cortex (basal) and associated with programs for automatic behavior like walking. While you are walking, your destination is determined in the cortex while the motor response is mediated largely by the basal ganglia.

Capgrass Syndrome Also known as the Capgrass delusion: a failure of a feeling of familiarity around people well known to the delusional patient. As a result, frequently encountered people are perceived as different, "imposters." The family dog may be declared to have been replaced by a duplicate. The dog looks and acts the same but there is something *different* about it

Cerebral Cortex The vast outer mantel of brain tissue covering the top and sides of the brain. It is involved in the storage and processing of sensory inputs and motor outputs, and contains the neurons.

Declarative Memory Memory that can be clearly expressed in words in response to an inquiry, such as "What is your address?" It includes semantic memory and episodic memory, both of which can be "declared."

Echolaic Memory The capacity to hold or recall a sound in one's mind. This is a form of sensory memory (auditory).

Emotional Binding Memories with powerful associations are more strongly encoded and easiest remembered.

Episodic Memory Memory for events that have been personally experienced at a particular time and location.

Hippocampus A seahorse-shaped structure connected to the amygdala that serves as the entry point for the formation of memory. The hippocampus is directly connected with nerve fibers conveying inputs from all of our senses (touch, vision, sound, taste, smell). It is thought to coordinate all of these sensory systems into episodic and long term memory. It is also connected with portions of the limbic system, principally the amygdala. You can visualize the hippocampus and the amygdala by closing your hand into a fist. The extended thumb represents the lower temporal lobe. When you move it outward, the tip of the thumb points towards you and the position of the amygdala and the hippocampus can be visualized as occupying the curved upper margin of the index finger, as it presses inward against the thumb area of the palm.

Iconic Memory The capacity to hold or recall a visual image in the mind. This is a form of sensory memory (visual).

Implicit Memory A form of non-declarative memory encoded by repeated experiences, but without any conscious attempt to learn. Closely allied to procedural memory since it doesn't involve consciousness.

Limbic System The mediator of all things emotional consisting of interconnected regions at the center of the brain, deep below the cortex. It includes on both sides of the brain the temporal lobe, the thalamus and hypothalamus, the cingulate cortex, the hippocampus, the amygdala, and other areas in the immediate surroundings.

Motor Program The sequence of movements that can be carried out automatically with little or no attention.

Multi Coding Linking the information to be learned with two or more sensations (usually visual and auditory). It is a key element of most mnemonic systems.

Non-Declarative Memory Employs no conscious expression and includes procedural or implicit memory "Doing and knowing without saying."

Perceptual Fluency The feeling of familiarity resulting from the smooth integration of present and past experience.

Semantic Memory General knowledge learned through repetition. In most cases the precise origin of the semantic memory cannot be consciously recalled. You remember that Dwight Eisenhower was present immediately prior to JFK, but you can't recall the exact occasion when you learned that fact.

Sensory Memory Brief transient sensations of immediate perceptions as when a person is actively seeing, hearing or tasting something.

State-Dependent Memory The likelihood of remembering something is increased by recreating the context of the original learning experience in terms of place, time, mood, etc.

Working Memory A mental workspace where short term memory is retained "online" and can be rearranged and manipulated within the mind into different configurations, depending on one's intentions at the time.

BIBLIOGRAPHY

PAPERS & JOURNAL ARTICLES

Cousins, J. N., K. F. Wong, B. L. Raghunath, et al. "The long term memory benefits of a daytime nap compared with cramming." *Sleep,* 2019: 42 (1): zsy207.

Dennis, D., A. C. Shapiro, et al. "The roles of item exposure and visualization success in the consolidation of memories across wake and sleep." *Learning & Memory* (2020): 451–456.

Lahl, O., C. Wispel, G. Willogens, et al. "An ultra-short episode of sleep sufficient to promote declarative memory performance. "*J. Sleep Res.* 2008; 17: 3-10.

Leong, R. L. F., N. Yu, J. L. Ong, et al. "Memory performance following napping in habitual and non-habitual nappers." *Sleep* 2021; zsaa 27.

Ong, J. L., T. Y. Lau, X. K. Lee, et al. "A daytime nap restores hippocampal function and improves declarative learning." *Sleep* 2021.

Van Schalkwijk, F. J., C. Sauter, K. Hoedlmoser, et al. "The effects of daytime napping and full night sleep on the consolidation of declarative and procedural information." *J. Sleep Res.* 2019; 28 (1):e12649.

Verhaeghen, Paul. "People Can Boost Their Working Memory Through Practice." The American Psychological Association's *Journal Of Experimental Psychology: Learning, Memory And Cognition,* Volume 30, No.6; (Nov. 4, 2004).

BOOKS

Abrams, Nelson Delis. *Remember It*! New York: Image, 2018.

Baddeley, Allan D. *Essentials of Human Memory*. Psychology Press Limited, 2006.

Carruthers, Mary, and Jan M. Ziolkowski, eds. *The Medieval Craft of Memory: An Anthology of Text and Pictures*. University Of Pennsylvania Press, 2002.

Cicero. *Rhetorica AD Herennium*. Translated by Harry Caplan. Harvard University Press, 1954.

Critchley, Simon. *Memory Theater*. Other Press, 2014.

Foer, Joshua. *Moonwalking with Einstein: The Art and Science of Remembering Everything*. New York: The Penguin Press, 2011.

Gluck, Mark A., Eduardo Marcado, and Catherine E. Myers. *Learning and Memory: From Brain to Behavior.* New York: Worth Publishers, 2008.

Joordens, Steve. *Memory and the Human Lifespan.* The Great Courses.

Lorayne, Harry, and Jerry Lucas. *The Memory Book*. Stein and Day Pub, 1974.

McGaugh, James L. Memory and Emotion: *The Making of Lasting Memories*. Columbia University Press, 2003.

Nu, Alain. *State of Mind*. C.F. B Productions Inc., 2015.

Papanicolaou, Andrew C. *The Amnesias: Clinical Textbook of Memory Disorders*. Oxford University Press, 2006.

Pribram, Karl H. *The Form Within: My Point of View*. Prospecta Press, 2013.

Roediger, Henry III, Yadin Dudai, and Susan M. Fitzpatrick. *Science of Memory: Concepts*. Oxford University Press, 2007.

Tolving, Endel, and Fergus L. M. Craik, eds. *The Oxford Handbook of Memory*. Oxford University Press, 2000.

White, Ron. *The Military Memory Man*. 2009.

Yates, Frances A. *The Art of Memory*. Penguin Books, 1978.

ACKNOWLEDGMENTS

Christopher Baldassano, Columbia University
Anna Shapiro, University of Pennsylvania
Mark A. Gluck, Rutgers University-Newark
Martha J. Farah, University of Pennsylvania
Steve Joordens, University of Toronto Scarborough
Barry Gordon, Johns Hopkins University
Tony Dottino, founder United States Memory Championship
Michael Dottino, developer Brain Power Series
Giuseppe Aquila, CEO Montegrappa
Nelson Dellis, four-time United States Memory Champion
Alain Nu, mentalist and mnemonist
Jim Karol, mneumonist.
Special thanks to Denise Leary for her advice and Franziska Bening
for her help in keeping this project on schedule.